Mission Mathematics II

Grades 6–8

Edited by

Michael C. Hynes
Juli K. Dixon

Contributors

Conni Blomberg
Michael Endress
John Kramer
Larry Lebofsky
Donna J. Moss
Vincent O'Connor
David F. Ortega
James E. Pratt
Deborah A. Shearer
Brad R. Thode

Project Director

Michael C. Hynes

NCTM
NATIONAL COUNCIL OF
TEACHERS OF MATHEMATICS

Copyright ©2005 by
THE NATIONAL COUNCIL OF TEACHERS OF MATHEMATICS, INC.
1906 Association Drive, Reston, VA 20191–1502
All rights reserved

Library of Congress Cataloging-in-Publication Data

Mission mathematics II : grades 6–8 / edited by Michael C. Hynes, Juli K. Dixon; contributors, Conni Blomberg ... [et al.].
p. cm.
Includes bibliographical references.
ISBN 0-87353-572-3
1. Mathematics—Study and teaching (Elementary) I. Title: Mission mathematics 2. II. Hynes, Michael C., 1941– III. Dixon, Juli K. IV. Blomberg, Conni.
QA135.6.M57 2004
510'.71'2—dc22

2004019027

The National Council of Teachers of Mathematics is a public voice of mathematics education, providing vision, leadership, and professional development to support teachers in ensuring mathematics learning of the highest quality for all students.

The publications of the National Council of Teachers of Mathematics present a variety of viewpoints. The views expressed or implied in this publication, unless otherwise noted, should not be interpreted as official positions of the Council.

Cover photograph: Nebulas Surrounding Wolf-Rayet Binary BAT99-49.
©ESO, European Southern Observatory/Y. Nazé, G. Rauw, J. Manfroid, J. Vreux, Y. Chu. All rights reserved. Used with permission.

Text photographs, unless otherwise credited, are courtesy of NASA.

PRINTED IN THE UNITED STATES OF AMERICA

Dedication

Mission Mathematics II is dedicated to the memory of Pamela Louise Mountjoy. Her passion for NASA, mathematics, and teaching served as the inspiration for Mission Mathematics II and has promoted the expectation that this series will make a valuable contribution to the lives of the students and teachers in our great nation.

Contents

Foreword

MISSION Mathematics: Linking Aerospace and the NCTM Standards is a collaborative project of the National Aeronautics and Space Administration (NASA) and the National Council of Teachers of Mathematics (NCTM). The vision of Frank Owens, director of educational programs for NASA, and James Gates, executive director of NCTM at the time of the conception of this project, inspired this unprecedented effort to link the science of aeronautics with the standards NCTM has developed for all aspects of mathematics education.

The books of the Mission Mathematics series have been well received by teachers year after year since their initial publication. However, both NCTM and NASA have experienced exciting changes, and the time for updating the series has arrived. The revision of Mission Mathematics has focused on aligning the activities with NCTM's *Principles and Standards for School Mathematics* (2000). To accomplish this alignment, the series now includes a new book for middle school teachers.

The middle school book is the bridge between the grades 3–5 book and the high school book, and in that sense, it mirrors the reality of middle school teaching. Middle school teachers continually strive to capture the best of two worlds: the nurturing, collaborative environment of middle school education and the subject-matter-oriented high school curriculum. This new book includes current information about the work and findings of NASA, and new and revised activities with additional resources to enhance all activities.

The original writing-team members, Vincent O'Connor, Conni Blomberg, Michael Endress, John Kramer, Donna Moss, David Ortega, James Pratt, Deborah Shearer, Brad Thode, Nancy Belsky, Catherine Blair, Eva Farley, Donn Hicks, Mary Ellen Hynes, Molly Ketterer, Barbara R. Morgan, Andrea Prejean, and Terry Thode, gave many, many hours to this project. Cyndy Rosso and Marilyn Hala, both NCTM staff members during the first-edition working period, contributed a great deal to the success of the original middle school book. Their support and expertise have added immensely to the quality of this product. The editors are grateful for their hard work, creative ideas, enthusiasm, and support throughout the project.

A special note of thanks is offered to Harry Tunis and the production and editorial staff of NCTM and posthumously to Pamela Mountjoy of NASA. Each provided the support, guidance, and expertise that have made the project enjoyable from the first day.

Clearly, this book was the product of many hands and minds. The collaborative efforts of many people made this important project a reality. Thank you all.

Michael C. Hynes, Editor
Juli K. Dixon, Associate Editor

Preface

THE goal of the writing team for *Mission Mathematics II: Grades 6–8* was to design mathematical problems and tasks that reflect NCTM's *Principles and Standards for School Mathematics* (2000) in the context of aerospace activities. Further, the writing team attempted to create activities that focus on—

- actively engaging students in NCTM's Process Standards: Problem Solving, Mathematical Reasoning and Proof, Communication, Representation, and Connections (among topics in mathematics, with other disciplines, and with real life);
- translating the work of engineers and scientists at NASA into language and experiences appropriate for young adolescents; and
- providing teachers with mathematics activities that complement many of the available NASA resources for students and educators.

The development of many activities for this project included pilot tests in classrooms throughout the nation. For the pilot, classrooms were selected to give a broad geographical, ethnic, and socioeconomic view of the effectiveness of the lessons. The large number of volunteers afforded the opportunity to reach a wide range of students. For those teachers who participated by sending feedback from their classroom experience with draft activities, thank you. Your feedback allowed the writing teams to make the activities more effective for students.

NASA is organized to implement the National Space Exploration Vision (www.nasa.gov). This book was developed with the collaboration of staff from NASA Education. The activities are focused on NASA's missions and research. Teachers can coordinate these mathematics activities with the many free materials available at the NASA Educator Resource Centers or NASA CORE (www.nasa.gov/audience/foreducators/CORE.html).

- What design features are built into these materials?

Thematic Modules

Many of the middle-grades materials are built around investigations into NASA's research pursuits. Thus, instead of being a collection of independent, one-day lessons, these materials consist of thematic units that may span several weeks. In line with curriculum planning for many middle schools, these modules afford multiple opportunities to connect mathematics with other disciplines. Many of the lessons, however, can be used independently. Some modules have multiple activities that may require more than one class period.

Discourse

The *Standards* documents describe a learning environment that promotes rich discourse among students. *Mission Mathematics II: Grades 6–8* was designed to optimize student-to-student discourse through the formation of student teams that engage in discovery, decision making, and problem solving. Numerous opportunities are also provided for teams to develop and present their findings in debriefing sessions to the entire class.

Assessment

In keeping with the vision of NCTM Standards documents, the modules in *Mission Mathematics II: Grades 6–8* offer multiple opportunities for teachers to monitor the progress of their students through work products, observation, logbooks or journals, and discourse. Students are encouraged to make daily entries into logbooks to document their experiences during their "missions." In most modules, students complete other work products that provide tangible evidence of their achievements.

The Formation of Student Teams

Most of the activities in this book require the formation of student teams. One reason for this approach is to maximize the active participation of all students in the learning process. A second reason—perhaps more important to the authors—is to be true to the way that NASA does business. Cooperative teams are the hallmark of virtually every successful business today. NASA is no exception and may even be a model of how teams get involved at every level in improving the chances for success on their missions. Middle school students are developmentally ready for teaming in small groups, but they may not have had the experience of working cooperatively with their peers. Given that traditional textbooks are designed to fit the model of individual learners working alone, they may not lend themselves to introducing and developing cooperative learning skills with middle school students. *Mission Mathematics II: Grades 6–8* modules may be especially effective in helping the middle school teacher form and promote the development of cooperative teams.

Mathematical Content

Each lesson in *Mission Mathematics II: Grades 6–8* correlates with the Content Standards for grades 6–8 in *Principles and Standards for School Mathematics* (NCTM 2000). Most teachers, however, will use these materials in conjunction with a basal textbook series. For this reason, *Mission Mathematics II* does not attempt to duplicate the expository or practice materials that are usually included in basal textbooks. Rather, the activities in this book are designed to be blended with other sources to give students additional explanations of ideas or practice with skills that are introduced or used in these activities.

✦ How do I use these materials?

Mission Mathematics II modules and activities naturally motivate middle school students to study mathematics. Learning in a meaningful context is important for all students and perhaps more so for middle school students. The activities in this book lend themselves to a variety of classroom approaches.

— The activities can be conducted independently to show the uses of mathematics in the space industry. Many of the activities are linked to form in-depth investigations that can be used in flexible ways.

— The investigations can become replacement units for parts of the current school curriculum. If teachers wish to use the materials in this way, the whole unit can be done at once or the activities can be conducted at regular intervals (for example, every Monday) as standard units of study in the mathematics curriculum.

— These investigations are ideal for schools that have instituted any form of block scheduling in mathematics. The longer class periods allow for active participation in activities, group work, data gathering, and journal writing.

Regardless of the organization of the school or the classroom, the modules, activities, and investigations in *Mission Mathematics II: Grades 6–8* will prove useful in helping middle school students learn and appreciate mathematics.

An image taken by the Hubble Space Telescope (HST) on February 22, 2002, shows a strange object that looks, in an uncanny way, like a hamburger. Arturo Gomez, an astronomer from the Cerro Tololo Inter-American Observatory in Chile, discovered this *proto-planetary nebula* in the constellation Sagittarius as he studied HST images. This shape is composed of the shadow of a thick disk of dust "sandwiched" between two other layers of dust that are reflecting light from a star.

Our Solar System and Beyond . . .

Functioning successfully in our world involves being aware of the immediate neighborhood, understanding its location in the city, realizing that the local neighborhood interrelates with other neighborhoods and cities, and so on. If we think of our neighborhood as being our entire planet, our horizons are expanded to a point where we probably will not be able to experience other such neighborhoods in person. The National Aeronautics and Space Administration (NASA) is dedicated to studying "neighborhoods" in space. Among the questions that scientists seek to answer in the study of space science are these:

- What is the origin of the Sun, our Earth, and the rest of the solar system?
- Do other stars have planets?
- What is the universe? How did it come into being? How does it work? What will be its ultimate fate?

From years and years of studying the heavens with ground-based telescopes, we have gained knowledge of the universe. We know that our star, the Sun, is one of perhaps a billion stars in our galaxy, the Milky Way. The Milky Way is one of a billion galaxies in the universe. These data make our space neighborhood seem very crowded. Billions of stars is an enormous number.

Furthermore, we know that Proxima Centauri, the star nearest to the Sun, is at a distance of only 4.3 light-years from the Sun. The nearest galaxy to the solar system, the Andromeda Galaxy, is 2.3 million light-years from Earth. Distant galaxies may be as far as 10 billion light-years away. Indeed, the universe is very large, and great distances separate stars and galaxies.

The ability of scientists to expand our knowledge about the universe has been hampered by Earth's atmosphere. The weather and the generally hazy atmosphere of Earth often distort and dim the images of ground-based telescopes. Scientists have counteracted this problem by placing telescopes in higher and higher places in the world. However, the Space Shuttle, with its ability to carry heavy loads, has enabled the placement of a telescope in orbit high above Earth's atmosphere. The Hubble Space Telescope (HST) was placed in orbit by the Space Shuttle in 1990. The HST weighs 11,600 kilograms. It is 13 meters long and

Teacher Research

The "Space Science Education Resource Directory" is a convenient way to find additional NASA space-science resources for classroom use.

http://teachspace.stsci.edu/cgi-bin/ssrtop.plex

Astronauts making repairs to the Hubble Space Telescope

about 4 meters in diameter at its widest point. It is about the size of a railroad tank car.

The HST has allowed scientists to make many discoveries since it became operational. Scientists now have—

- the first conclusive evidence of the existence of black holes;
- indications that the universe may be much younger than previously thought;
- the first direct evidence that the universe is evolving;
- information that quasars do not dwell in cores of galaxies but are isolated in space; and
- indications that dark matter in the universe is more exotic than once thought.

Through the continuing research efforts of NASA scientists and their collaborating colleagues and the use of the high technology of the space program, we are learning much about the universe. Our knowledge of our neighborhood is growing exponentially.

Learning about the Solar System

Most middle school students have some knowledge of the solar system. They know that Earth has a moon and revolves around the Sun. They probably can name several planets and are familiar with Mars and Saturn. Some students will have learned a mnemonic device for remembering the order of the planets from the Sun to Pluto, such as "My Very Earnest Mother Just Sat Upon Nine Pins."

Probably only *Star Trek* fans or budding astronomers know that this order is not constant. At times, Uranus is the outermost planet, not Pluto. Because the orbits of these two planets are more elliptical, Pluto and Uranus are sometimes configured so that Pluto is actually closer to the Sun.

This module contains four activities. These activities can involve between ten and fifteen class periods; the time spent depends on how many optional portions of the activities, extensions, and assessments are used. Some portions of the activities are connected, but even with this connected design, the later activities can be taught separately as long as the necessary materials produced through the earlier activities are prepared in advance.

- "Orbit Lab": In this introductory activity, students explore some rudimentary properties of orbits. Through the use of manipulatives, students experience some of the properties of orbits and think about the mathematics associated with the properties.
- "A Planetary System": Students begin exploring the characteristics of orbits, including that of an imaginary planetary system, Delphi. In this part of the module, students use concepts from their study of fractions and number theory to predict the occurrence of planetary alignment.
- "Our Solar System": Students gather data about our solar system. They use library resources, including NASA publications, to create a class chart or poster containing information about the planets, their moons, and the Sun. Using the chart of data about our solar system, students create a model showing the distances of the planets from the Sun. Proportions are used to scale the distances among the orbits. In this first attempt to build a model of the solar system, students use model planets that are not proportional with respect to size.
- "1000-Yard Solar System": With some experience trying to build a model of the solar system, students have begun to conceptualize its immense size. Students are now ready to create a model in which both the planets' distances from the Sun and the sizes of the

A view of Earth from the Moon

planets conform to the same scale. Proportional reasoning is used to establish the distances in the model, as well as the sizes of the planets. The culmination of the activity involves taking measurements for the model and creating it outdoors.

The outdoor portion of the "1000-Yard Solar System" activity requires advance preparations and clearances from the school administration. Please read the suggestions in the activity description.

Journal writing is an important part of learning mathematics. By formulating their thoughts about their understanding of mathematics related to space science, students are completing an essential step in the learning cycle—reflection. Completing the mission logbook is a significant part of this module. A sample log sheet is included on page A-2 in the appendix.

Developing a mission patch is another important part of the module and of actual Shuttle missions. Each Shuttle crew designs a mission patch to represent the objectives of the mission and the characteristics of the crew. The design of the patch is taken seriously and is often started as soon as the team begins working together. Each mission patch is a permanent fixture in the history of NASA space flight. As students create art, they often are more capable of reflecting on what they have learned. Developing mission patches that show both the space-science mission and the related mathematics allows students to engage in important reflective thinking. The patches can be used as logbook-cover designs. Illustrating the cover of the logbook can become an ongoing assignment as other Mission Mathematics modules and activities are completed.

ACTIVITY 1

Orbit Lab

In this first activity, students explore concepts related to orbits. Although students may have learned that planets are in orbit around the Sun, they may not have had an opportunity to think about some of the basic principles of orbital mechanics. This activity involves the hands-on collection of data; for this reason, students should be grouped into mission teams of three to five members. Reinforce the idea that although flying in the Space Shuttle is one type of mission, many research activities in space science can also be called missions.

Important Mathematical Ideas

Students use geometry knowledge and measurement skills in a hands-on activity that requires them to make inferences from data and think about accuracy of measurement.

Mission

The introductory discussion helps students develop the vocabulary and concepts that are used in the subsequent activities of this unit. The laboratory experiences help the students construct their knowledge about orbital mechanics. Using a relatively constant speed for an orbiting planet, students explore the relationship between the size (radius) of an orbit and its period.

Materials and Equipment

Each group of students needs the following materials: Approximately six feet of string, six heavy washers, one rubber stopper, three inches of masking tape, a stopwatch, a meterstick, a calculator, data sheet 1 (see appendix, page A-3), a log sheet (see appendix, page A-2), and graph paper.

Launching the Activity

In the whole-class portion of the activity, brainstorming is used both to promote students' interest in the activity and to assess students' knowledge levels. If brainstorming is a new class activity, guidelines need to be set for participation. Students should be encouraged to give their ideas freely and openly, and all responses should be accepted as valid. Rejecting one student's idea may discourage another student from making a conjecture. Students' conjectures can be reviewed at a later time in the activity or near the end of the unit to determine

NCTM Standards

Instructional programs from prekindergarten through grade 12 should enable all students to—

Geometry

Analyze characteristics and properties of two- and three-dimensional shapes

Measurement

Apply appropriate techniques, tools, and formulas to determine measurements

Data Analysis and Probability

Develop and evaluate inferences based on data

Algebra

Represent and analyze mathematical situations and structures using algebraic symbols

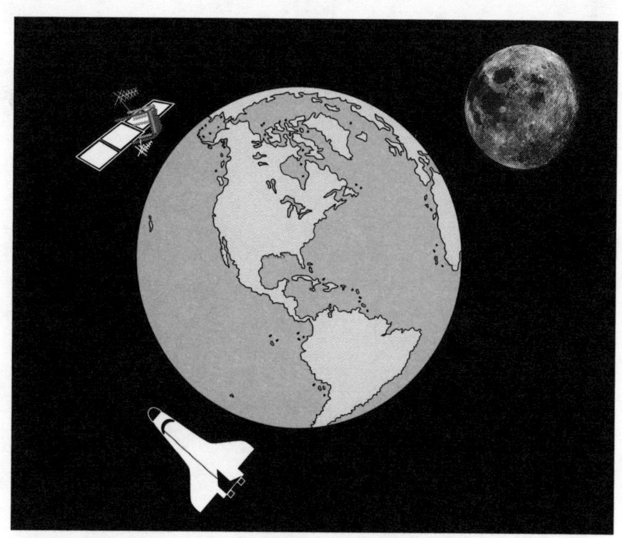

Weather satellites, Space Shuttles, and the Moon are just some of the many objects that orbit Earth.

which seem most valid. These ideas can be noted on a bulletin board or chart to create a class record for the unit.

The International Space Station

To initiate brainstorming with this activity, ask an open-ended question, such as "What are some examples of objects that are in orbit around Earth?"

Ideas should be elicited exclusively from the students. Great patience may be required to get the first response. The "wait time" for subsequent responses should be reduced. List all students' responses on chart paper or on the chalkboard. If the chalkboard is used, have a student record the list on paper for future use.

Do not be surprised if students voice misconceptions about astronomy. Many students have not yet constructed their understanding of the relationships among moons, planets, and stars. They may suggest that the Sun or planets are in orbit around Earth. At this point in the unit, these notions should be accepted as conjectures. As the activities of the unit progress, the list of conjectures can be revisited and revised to remove misconceptions.

The brainstorming session should indicate the students' understanding of orbits and basic astronomy. This information is valuable for helping to determine the amount of time needed to complete all of this activity. Note that this portion of the activity may require a whole class period. To close the brainstorming session, summarize the discussion.

Developing the Activity

Begin the laboratory part of the activity with a presentation of the formal concept of an orbit: *An object is considered to be in orbit if it is traveling in a circular or oval (elliptical) path around Earth or another object in space. The time required for an object to travel through one complete orbit, or revolution, is called the* period *of the orbit.*

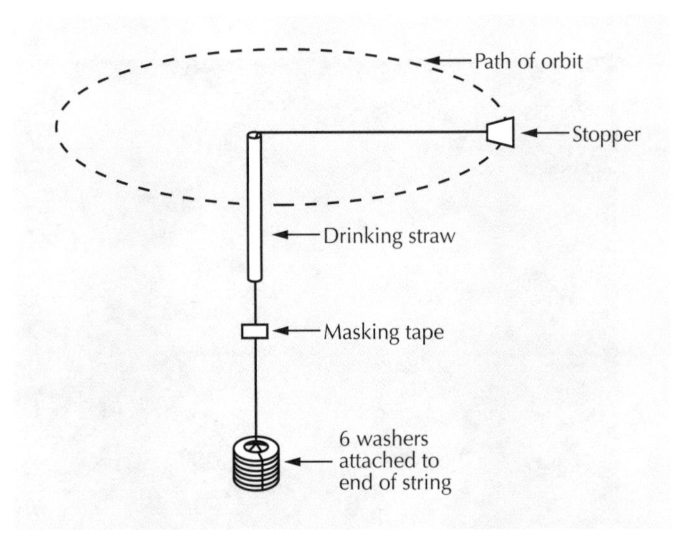

This part of the "Orbit Lab" activity is a laboratory exercise appropriate for small groups of students. Mission teams of three to five students should be formed. Roles for four students are presented below. These roles can be combined or shared for groups of three or five.

Spinner: Spins the orbit simulator at a uniform speed
Timer: Uses the stopwatch to record time
Counter: Counts the number of revolutions
Recorder: Records data

Each mission team needs to assemble the orbit simulator as shown at left.

Some students may not be familiar with the term *simulator.* Take time to discuss the notion of simulators as models of real-world settings. They are important for learning in laboratories, training in the workplace, and training in the military. The development of simulators has become an important and growing industry.

The distance from the bottom of the stopper to the straw on the assembled orbit simulator is the radius of the orbit. For trial 1, the radius should be 60 centimeters. Masking tape is placed on the string about 2 centimeters below the straw. The mission team's spinner holds the straw above the level of his or her head, keeping the masking tape in view. The student spins the simulator by moving his or her wrist in a circular motion. The spinner continues to spin the stopper in such a way that the masking tape stays approximately 2 centimeters from the bottom of the straw.

Hold the straw loosely enough to allow the string to slide easily up or down. Note that increasing or decreasing the speed causes the tape

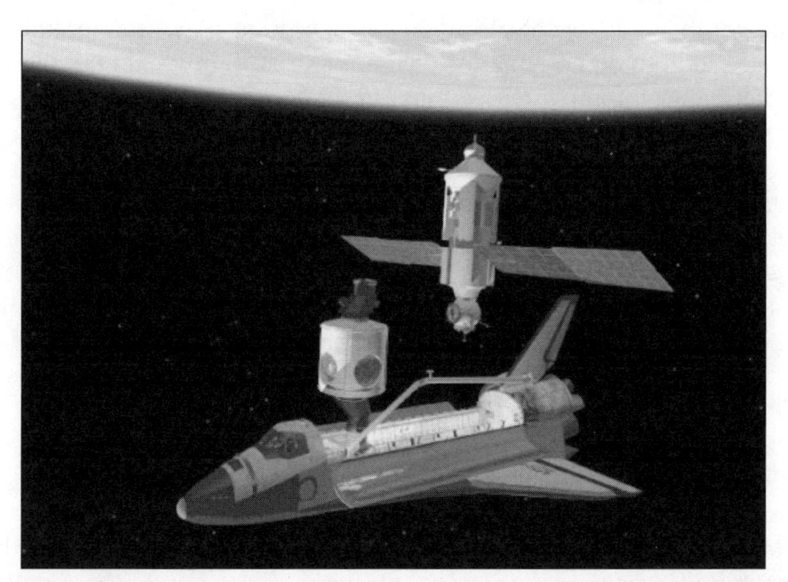

Picture one shows the shuttle and the International Space Station (ISS). Note the docking ring on the shuttle. The second picture shows the shuttle docked to ISS.

Safety Tips

Students should be warned against letting go of the spinning device. Caution students about fastening the cork securely to the string.

Teaching Tip

Students who are unfamiliar with stopwatches will find short durations of time difficult to measure. Thirty revolutions of the orbit simulator are recommended to improve the accuracy of the measurements. The number of revolutions can be adjusted according to the students' ability to use stopwatches.

Mars at opposition (2001)

to rise or fall. At a uniform speed, the tape location and the radius of the orbit remain constant.

Once the mission team's designated spinner has mastered the ability to keep the speed constant, the counter tells the timer to begin and counts thirty revolutions. The counter should count out loud so that the timer will be ready to stop at the end of thirty revolutions. The mission team's recorder enters the time on the data sheet. The data sheet for this laboratory exercise is found on page A-3 in the appendix.

Repeat the process for radii of 65, 70, 80, and 85 centimeters. Each student on the mission team should record the results in the first chart on the data sheet.

Each student plots the collected data on a graph, which should show radius on the horizontal axis and time on the vertical axis. Make certain that the axes are labeled. After the data are plotted, a discussion is necessary. At this point, students should sketch a curve through the points on the graph. Middle school students may tend to draw straight lines and may need to be encouraged to draw curves if their data fall into that pattern. Reinforce the idea that the students are sketching the line.

If students have not noticed, point out that the entry for a radius of 75 centimeters is missing. Have students predict the value of the missing data by using the graph. This time should be recorded on the data sheet. The predictions of all mission teams should be shared.

Because students usually make different predictions for 75 centimeters, the whole class should discuss the accuracy of the predictions they have made. The notion that 100 percent accuracy is unlikely should be one outcome of the discussion. Predictions using approximations are subject to measurement error. The idea is to reduce the error as much as possible. One way to accomplish this goal is for each mission team to collect and graph the data several times. If the data are consistent, then predictions become more acceptable.

Teaching Tip

Note that absolute value is used in the accuracy formula to avoid negative percents. This activity can present an ideal opportunity to discuss the notion of absolute value in more detail.

Another way to improve predictions is to use the predictions of all the mission teams. Use a line plot of the individual predictions of all students to determine a class prediction. One technique for forming the line plot is to draw a number line on the chalkboard and affix sticky notes containing the individual predictions at appropriate places. If statistics has been included in the school mathematics curriculum, take this opportunity to introduce or practice using stem-and-leaf plots or box plots as a means of looking at data to determine a class prediction.

Prediction is used in many research areas of aerospace science. For example, NASA scientists need to predict the location and speed of the space station to enable the Space Shuttle or another orbiter to dock successfully. The key to successful docking is to reduce the margin of error as much as possible. Pilots of orbiters make corrections during flight to modify the predictions and complete the docking. The goal is to eliminate the need for such corrections by improving predictions.

In many real-world applications, a predicted value that is within 10 percent of the experimental data would be considered a good prediction. Considering the accuracy of the measurements in this experiment, 10 percent would be acceptable. To determine the percent of error, an actual time is needed.

Calculation of accuracy:

$$\left| \frac{\text{predicted value - actual value}}{\text{actual value}} \right| \times 100 = \text{error (\%)}$$

The predicted value in the accuracy formula was determined as a class in the line-plot exercise. Thus, the predicted value is the agreed-on class prediction.

Next have students return to their mission teams, collect data for an orbit with a radius of 75 centimeters, and record the time. This experimental value will become the actual value in the accuracy formula. Students may use calculators to determine the accuracy of their predictions.

Calculation of accuracy in this experiment:

$$\left| \frac{\text{class prediction - experimental value}}{\text{experimental value}} \right| \times 100 = \text{error (\%)}$$

Concluding the Activity

Mission teams can repeat the experiment to test the accuracy of their new predictions. Have them use different lengths of string. Using longer strings makes counting the revolutions easier.

All students should take time to write in their logbooks about what they learned in this activity. The log sheet has guiding questions to assist students in their reflections.

Some students may want to begin thinking about their mission patch for the "Learning about the Solar System" unit.

Extending the Activity

Students can determine the period of the stopper orbits by dividing the time for thirty revolutions by 30. The radius versus the period can be graphed, and predictions can be made for the missing data.

Students can calculate the speed of the stopper using this formula:

circumference × number of revolutions / time.

Graphs can be made of time versus speed or speed versus period.

NASA Connection

Students who are ready to explore on the Internet may want to investigate the following sites:

www.jpl.nasa.gov/index.html

www.jpl.nasa.gov/solar_system/planets/mars_index.html

mars.jpl.nasa.gov

www.nationalgeographic.com/ngm/0102/feature2/index.html

heritage.stsci.edu

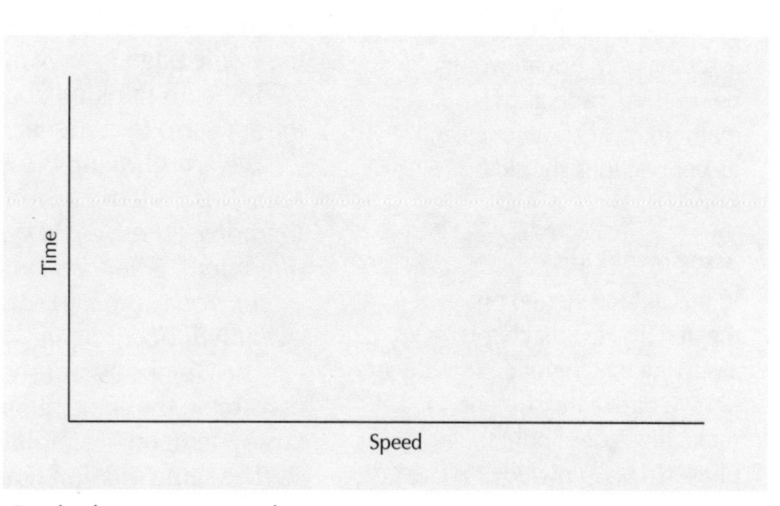

Graph of time versus speed

ACTIVITY 2

A Planetary System

NCTM Standards

Instructional programs from prekindergarten through grade 12 should enable all students to—

Geometry

Analyze characteristics and properties of two- and three-dimensional shapes

Number and Operations

Understand numbers, ways of representing numbers, relationships among numbers, and number systems

Understand meanings of operations and how they relate to one another

NCTM Assessment Standards

Student assessment should be aligned with, and integral to, instruction. Multiple sources of assessment information should be used. All aspects of mathematical knowledge and its connections should be assessed.

Assessment Tip

In this getting-started phase of the activity, the teacher is assessing the content knowledge of the students to make decisions about the next phases. This knowledge is used to develop the content and manage the class.

Important Mathematical Ideas

The purpose of this activity is to give students experience with least common multiple (LCM), a mathematical concept that is important in calculating with fractions and has many applications in the real world. The application to space science helps students understand the importance of LCM in a motivational context.

Mission

The goal of this activity is to determine the time required for planets to become aligned in orbit.

Through the research of NASA's scientists, we are beginning to identify planets in other star systems. In this activity, students study the planets of the solar system and think about the mathematics of an imaginary system of planets orbiting an imaginary star, Delphi. Students are challenged to solve a problem related to the alignment of planets in orbit around Delphi. This problem can be extended to our solar system.

Materials and Equipment

Five objects with one (the star) being larger, a piece of chart paper that can be used to draw four orbits around the star, a compass, and data sheet 2: "Planetary Revolution and Alignment in the Delphi System" (see appendix, page A-4).

Launching the Activity

Initiate the activity by asking the class what they know about the Hubble telescope and about what has been discovered through this orbiting telescope. Refer to the NASA home page on the World Wide Web if students want to learn about some of these discoveries. The address of the home page is www.nasa.gov.

To begin thinking about new planetary systems, present the following scenario to students:

We are studying the imaginary star system called the Delphi system. This system has only four planets. By some chance, the first photograph of the Delphi system showed that all four planets were in alignment. Scientists began to wonder how often this alignment would occur. Your task is to explore this idea and try to predict when this event will occur again.

The Delphi system should be presented as a system in which all planets orbit in the same plane. This example is not a true model of a planetary system, but it simplifies the situation for middle school students.

The same mission teams of three to five students formed in the "Orbit Lab" activity begin their work by naming the four Delphi planets. Once the planets are named, students are ready to begin studying their orbits. Record the names in table 1 on data sheet 2, which can be found on page A-4 in the appendix.

Developing the Activity

Begin with a simulation of the movement of the planets in their respective orbits by using manipulatives. Have students experiment with orbital paths by moving the "planet" manipulatives on paths drawn on chart paper. Have them speculate about which planet would take the least time to orbit the star. Which would take the longest time?

Orbital Periods of Delphi Planets

Planets	*Data Set 1*	*Data Set 2*
A	2	1
B	3	2
C	4	3
D	5	4

In this activity, we are concerned only with the periods of the Delphi planets. A *period* is defined as the time required for an orbiting body (planet) to complete one revolution around another body (star). This activity involves patterns among the given periods of the Delphi planets. Sample data sets follow, to be used with data sheet 2.

No unit of measure is associated with the orbital periods in the chart. Rather than use a year or some other familiar time unit, create a fictitious one called a *golly*. According to data set 1, planet A orbits Delphi in 2 gollies. Planet B completes a revolution in 3 gollies, and so on. Record these values in table 1 on data sheet 2.

Many middle school students need to simulate the planets' traveling in their respective orbits rather than deal with the problem abstractly. In these situations, a scaled value must be assigned to a golly. The time required for a planet in the Delphi system to complete an orbit may be many Earth days; for this reason, the simulation of the orbiting planets should use Earth seconds. Encourage students to choose a number of seconds to be used in the simulation. They should record this scaled value in table 1 on data sheet 2.

The tables on data sheet 2 will help students compute the LCM for the various orbits or simulate the orbital periods to determine the LCM pattern.

Instead of trying to solve the whole problem at once, students should be encouraged to consider subproblems. Each student should predict the length of time needed for planets A

Technology Link

If technology is available to students, use this opportunity to involve them with databases. Rather than make class charts of the data from the research effort, have students create a database of their discoveries.

Cards with names or pictures of the planets, moons, and the Sun

and B to come back into alignment, then record the prediction in table 2 on data sheet 2. Students should then simulate the revolutions, record the results in table 3 on data sheet 2, and compare the results with their predictions.

One way to simulate the realignment of two planets is to use one student to represent Delphi and two other students to represent planets. The "planets" should be in alignment to begin. Ask "planet A" to move in its orbit to the approximate position where it would be in 1 golly. If 5 seconds represents a golly, the student could walk slowly to the 1-golly position. The second student should be asked to move to the 1-golly position in the same manner. This movement should be repeated for 2 gollies, 3 gollies, and so on, until the "planets" are realigned. The result should be recorded in table 3 on data sheet 2 and compared with the prediction in table 2. This process should be extended to planets A, B, and C, then to all four planets. Ask students to look for patterns in the data.

If the students see no connection with LCM after simulating the first data set, a second simulation exercise can be done with data set 2.

Sometimes another approach is more fruitful than repeating a simulation. Table 4 on data sheet 2 offers another way to think about alignment and LCM. Here, students record a check in the column that represents a completed orbit for a planet. Table 4 is appropriate to use with data set 2.

Teaching Note

An important component of mathematical thinking is mathematical modeling. In the initial process of making mathematical models for real-world settings, mathematicians sometimes impose on a situation conditions that simplify the model. In this activity, some assumptions are made to allow middle school students to think about a relationship of orbital periods. Two such assumptions are that orbiting bodies travel at constant speeds and that the orbits are coplanar.

Teaching Tip

Some students may know that the period of a planet is directly dependent on the length of the orbit and the speed of the planet. Discussion of this relationship may become an extension of the activity.

Concluding the Activity

Explain to students that when mathematicians work with newly discovered concepts or notation, they may see no apparent connection with the real world. Often, they do not discover these connections until later. The LCM is one instance of a mathematics concept having seemingly little connection with the real world. But as can be seen, the LCM applies to orbital periods and simple notions of planetary alignment. It has other applications, as well. For instance, hot-dog buns are packaged with eight buns in a package, and hot dogs come ten to a pack. To make the number of buns and hot dogs come out even, use the LCM. Because the LCM has many real-life applications, have students write about other ways in which they think this concept applies in the real world.

Extending the Activity

"A Planetary System" focuses on an imaginary star and its planets. Encourage students to think about our solar system and the alignment of planets. Review the solar system information on data sheet 3 in the appendix (page A-5), and have students assume that the planets of our solar system are in alignment. Ask them to determine how many years will pass before the planets are again in alignment. Students used subproblems to analyze the imaginary system of Delphi. They can employ that same technique to begin aligning the planets of our solar system. Students should use technology to complete as much of the chart as they can.

ACTIVITY 3

Our Solar System

Important Mathematical Ideas

This activity is a first step in the development of students' understanding of the large numbers associated with distances in space. Students use ratio and proportion to help expand their number sense related to large numbers. Calculators help them build this understanding.

Mission

The goal of this activity is to create a classroom model of the solar system.

The space-science efforts of NASA have allowed us to improve our understanding of our neighboring planets. Since the 1960s, NASA scientists have been investigating Mars. Mariner missions 4, 6, and 7 were flyby operations that returned photographs and weather data from Mars. *Mariner* 9 revealed large mountains and deep valleys on the surface of Mars. The two Viking spacecraft that were sent to Mars in the 1970s gave us a better understanding of the Martian surface. The latest generation of Martian orbiting explorers, *Mars Pathfinder* and *Mars Global Surveyor,* will refine our knowledge even further. In 2003, the Mars Exploration Rover Mission sent two Mars rovers, *Opportunity* and *Spirit,* to the surface of the red planet. Early data indicate that at one time, water was, indeed, found on Mars. We are learning more and more about this relatively close neighbor and the other planets in our solar system. To better understand the planets, students need to develop an intuitive notion about the vastness of the solar system.

Materials and Equipment

A prepared data chart, "The Planets at a Glance," is provided in the appendix, (page A-6) as a resource for teachers. The class will also need calculators, string, and Styrofoam balls or circular cardboard cutouts to represent planets: at least two circles about 30 cm in diameter, at least two circles about 15 cm in diameter, at least two circles about 10 cm in diameter, and at least three circles about 8 cm in diameter. In the extension section, students use data sheet 4, "Calculating Proportional Distances for a Model of the Solar System" (see appendix, page A-7).

Launching the Activity

Conduct a whole-class discussion to make a list of the facts the class knows about our solar system. Ask questions involving, and pose tasks related to, facts about the solar system, such as those listed below.

- What are the relative sizes of the planets in the solar system?
- Order the planets according to the number of moons they have.

NCTM Standards

Instructional programs from prekindergarten through grade 12 should enable all students to—

Geometry

Analyze characteristics and properties of two- and three-dimensional shapes

Measurement

Apply appropriate techniques, tools, and formulas to determine measurements

Saturn

Jupiter

Uranus

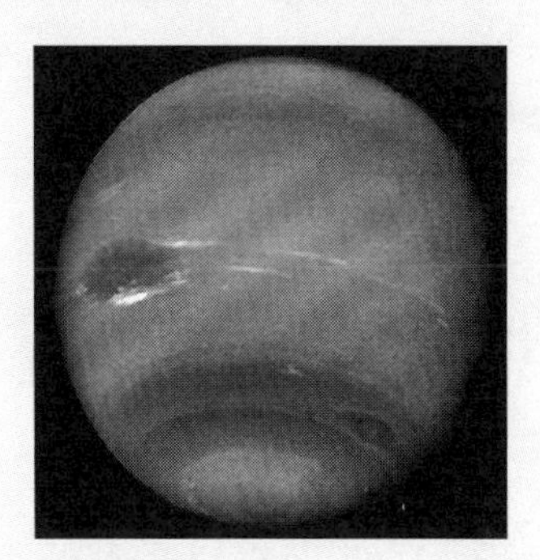

Neptune

Earth

Venus

Mars

Mercury

Pluto

- Order the planets by size.
- Order the planets according to their distances from the Sun.

During the discussion, determine students' understanding of the large numbers involved in astronomy. Ask students how they think about the size of these numbers. Determine how they comprehend the relative magnitudes of sizes and distances.

Information about students' understanding gained by listening and observing during the discussion can be used in making decisions about grouping students later in the activity. For example, during the research part of the activity, students who have high levels of knowledge should not be paired with those who have little knowledge. This type of grouping may result in little learning for the students with less knowledge because they tend to rely on the students who may have learned about the solar system before this class discussion. If groups are formed according to students' levels of knowledge about the solar system, all students will learn more in this experience.

Develop a list of information that each research group should find about its topic. Students should brainstorm to make the list. All data that interest students are important, but be certain to guide the discussion so that the following topics appear on the list:

- The size of the planet
- Facts about the orbit of the planet
- The mean distance from the Sun
- The number of moons

The information needed to complete the task can be found in various sources. NASA offers resource materials on planets and the solar system that can be placed in the school library. (See the resources listed in the bibliography [pages A-33–A-34].) Science books and encyclopedias have many of the desired facts. Students can also use the Internet to explore NASA's Web pages.

Work together as a class to create a chart or poster of all the data collected by students. The whole-group work that results in the poster or chart is essential to the mathematics of this activity. Students should also make written or oral presentations, or both, of their research. From these reports, important data can be added to the class chart or poster.

As the class summary is developed, students' attention should be focused on the magnitude and relative magnitude of the measures associated with the planets. Through research and discussions, students should have improved their capabilities to understand and compare large numbers.

Developing the Activity

Using the class chart or poster, conduct a discussion of the data about planets in the solar system. Because of the immense distances and sizes involved, middle school students may have difficulty conceptualizing the magnitude of the Sun and the planets and the distances between them. This discussion should relate the sizes and distances to the students' level of number sense for large numbers.

One way to increase students' understanding is to build scale models of the solar system. This activity presents the first opportunity to use this strategy. Although only estimations of ratios are employed in this activity, students begin to think about the magnitude of the solar system.

Challenge students to think about the relative sizes of the planets. Show them a set of Styrofoam balls or circular cardboard

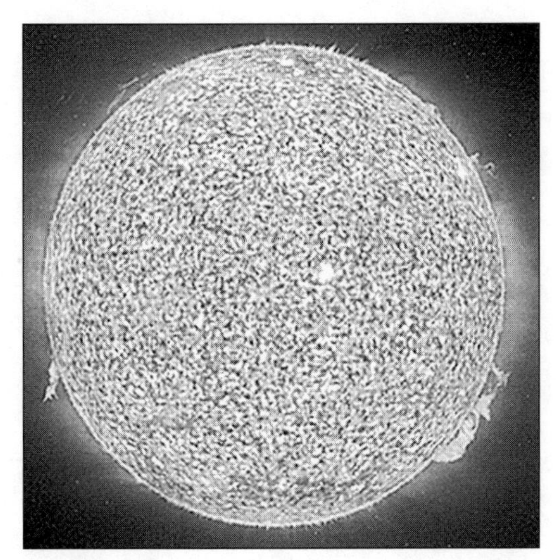

The Sun

NCTM Teaching Standards

The teacher of mathematics should orchestrate discourse by ... deciding when to provide information, when to clarify an issue, when to model, when to lead, and when to let a student struggle with a difficulty [and by] monitoring students' participation in discussions and deciding when and how to encourage each student to participate.

(NCTM 1991, p. 35)

NCTM Teaching Standards

The teacher of mathematics should promote classroom discourse in which students—

...

use a variety of tools to reason, make connections, solve problems, and communicate; [and]

try to convince themselves and one another of the validity of particular representations, solutions, conjectures, and answers.

(NCTM 1991, p. 45)

cutouts that could model the planets. Emphasize that these objects, the "planets," are not sized exactly to scale; they represent only four categories of the sizes of planets.

Show students an object of one size, and ask, "Which planets could be represented by a circle [or ball] of this size?" Ask the same question for each size. The resulting categories should have two very large planets—Saturn and Jupiter, two large planets—Uranus and Neptune, two medium planets—Earth and Venus, and three small planets—Mars, Mercury, and Pluto. Label a ball or circle for each planet, and give each "planet" to a student to hold.

After all the planets have been distributed, have the students holding the planets stand in front of the class. Ask them to order the planets by size from smallest to largest and to name the planets. Then ask them to arrange themselves in the order of the planets from the Sun as they appear in the solar system.

Begin a discussion of the size of the solar system. Have each student with a planet select another student to hold it. Tell this new set of students that the Sun is in one corner of the room. Ask, "Where do you think Earth is located?" Have the class decide where the student holding Earth should stand in the classroom. Continue by having the class place the other "planets" in correct order and in the locations they think the planets would be in the solar system. The students are now arranged in the correct order, but the distances from the Sun are most likely significantly out of proportion. In many classes, students do this activity without thinking about the data on the class chart. If this situation occurs, refer to the class data chart and discuss the relative distances between the planets. Repeat the placement of the planets, giving more attention to the spacing.

Concluding the Activity

Have the class make a hanging display with the Styrofoam balls or cardboard cutouts to represent the order of the planets in the solar system. If enough materials are available, groups of students can make displays.

Ask students to think about the distances in the solar system and write their understanding of these distances in their logbooks. Students can be encouraged to interview parents, siblings, or other adults about the size of the solar system and to record their ideas.

Extending the Activity

Earlier in this activity, the relative distances of the planets from the Sun were used to foster students' understanding of the solar system. In this part of the activity, students begin to use ratio to develop a reasonably accurate scale representation of the distances of the planets from the Sun.

To present the pertinent data for this activity, see data sheet 4 in the appendix (page A-7), and have students complete this data sheet.

Begin with a discussion of scale drawings and scale models. Ask students to give examples of scale drawings or scale models that they may have seen. Students may mention that a map has a scale. Ask, "What does it mean to have a scale?" Ask students questions about making a scale drawing of an automobile or another familiar object on a letter-sized piece of paper. Ask such questions as "If the automobile is 15 feet long, can we use a scale of 1 inch equals 1 foot?" Once students show an understanding of the use of ratio to create a scale, return to the problem of creating a scale model of the solar system.

Tell students that you want them to develop a scale model of the solar system for the classroom. This model should meet these criteria:

- All planets must fit inside the classroom.
- The planets must be far enough apart that the circles representing them do not overlap.
- All planets should be easily visible in the classroom.

In a typical classroom, these requirements cannot easily be met. However, the process of trying to develop the classroom scale shows students that the first attempt at solving a problem does not always lead to the best solution.

Begin the activity by assigning each student a planet other than Pluto. Tell students that the Sun is in one corner of the room and that Pluto is at the opposite end of the room, farthest from the Sun. Measure this classroom distance, and have students use it to develop the scale for the model with the help of their calculators. Use the mean distance from the Sun to Pluto on data sheet 4 and the distance across the classroom to set up the ratio for the scale.

Suppose that the distance between the classroom Sun and the classroom Pluto is 40 feet. Use this value to set up a ratio with the actual mean distance from Pluto to the Sun:

The actual mean distance from Pluto to the Sun is 3,672,000,000 miles.

The mean distance from the classroom Pluto to the classroom Sun is 40 feet.

1. Convert miles to feet so that both measures will be described using the same unit. Because a mile is 5,280 feet, the distance from Pluto to the Sun is multiplied by 5,280 feet/mile:

   $$3{,}672{,}000{,}000 \text{ miles} \times 5{,}280 \text{ feet/mile} = 19{,}388{,}160{,}000{,}000 \text{ feet}$$

2. We have a ratio of

   $$\frac{40 \text{ feet}}{19{,}388{,}160{,}000{,}000 \text{ feet}}.$$

That is, 40 feet in the classroom model represents 19,388,160,000,000 feet in real life.

Determine the distance from the classroom Earth to the classroom Sun.

The actual mean distance from Earth to the Sun is 93,000,000 miles.

The mean distance from the classroom Sun is unknown. Convert the unit for the actual mean distance from miles to feet.

NCTM Teaching Standards

The teacher of mathematics should pose tasks ... that ... call for problem formulation, problem solving, and mathematical reasoning.

(NCTM 1991, p. 25)

NCTM Mathematics Curriculum and Evaluation Standards

Students should be able to select and use tools appropriate to solve a problem.

(NCTM 1989, p. 8)

Teaching Tip

This activity involves computation with long distances. Because most calculators have eight-digit displays, students will need to use number sense to determine the proper place value for the distances. If scientific notation has been part of the curriculum, this activity offers an appropriate opportunity to practice the use of this notation.

NCTM Assessment Standards

[Assessment] is an integral part of instruction that encourages and supports further learning.

(NCTM 1995, p. 13)

1. $93{,}000{,}000 \text{ mi} \times 5280 \text{ ft/mi} = 491{,}040{,}000{,}000 \text{ ft}$
2. The result, then, is a ratio of

$$\frac{n}{419{,}040{,}000{,}000},$$

in which n = the classroom model of the mean distance from the Sun in feet. The proportion using the classroom and actual distances between Pluto and the Sun and the classroom and actual distances between Earth and the Sun can be computed with the following equation:

$$\begin{aligned} \frac{40}{19{,}388{,}160{,}000{,}000} &= \frac{x}{491{,}040{,}000{,}000}, \\ 19{,}388{,}160{,}000{,}000x &= 40 \times 491{,}040{,}000{,}000, \\ 19{,}388{,}160{,}000{,}000x &= 19{,}641{,}600{,}000{,}000, \\ x &= 1.013071895, \end{aligned}$$

or n equals approximately 1.013 feet. Thus, the scale-model distance of the classroom Earth from the classroom Sun is about 1 foot.

Have each student use a calculator and the scale to determine the classroom distance of his or her planet from the Sun. Select a planet, and have all students assigned to this planet give their calculations of the distance. Take advantage of this opportunity to check students' use of the calculator to work out these scaled numbers. Select students to make the agreed-on scaled measurement, and place the planet in the proper location. Continue until all planets have been placed in the classroom model.

When the model is complete, have the class evaluate it using the criteria established earlier in the activity. If the classroom is like most, the planets close to the Sun will be small distances apart. The circles representing some of these smaller planets may even overlap, an outcome that clearly violates one of the criteria.

This situation can be used to initiate more discussion about these distances. Students will comment on how close some planets are in the model. Ask, "Is this overlap caused by errors in measuring or calculating the scaled distances?" The actual planets do not overlap. The choice of the scale has made a model in which the planets near the Sun seem extremely close together; however, this situation is not realistic. The effort to contain the model in the classroom has resulted in a model that is not the most desirable.

If the school building has hallways, students can repeat the development of the model by placing the Sun and Pluto at the opposite ends of a hallway. Some teachers have used a corner of the classroom for the Sun and allowed the model to extend out the classroom door. Placing Pluto outside the classroom can stimulate students in other classes to begin thinking about space science.

Concluding the Activity

Students do not necessarily understand the scale of space after just one activity; achieving this awareness is a journey. After this classroom activity, ask students to reflect on how their understanding of space has been altered and to write their thoughts in their logbooks. Also, have students write about the mathematics they used in the activity, as well as the role that the calculator played in helping them solve the problems.

If some students finish their reflection activities quickly, they can begin developing their mission patches.

Teaching Note

Making a true scale model of the solar system in a limited area is not practical. The distances are too great. In the model used in this activity, the scale used for the distances from the Sun is different from that used for the sizes of the planets.

ACTIVITY 4

1000-Yard Solar System

NCTM Standards

Instructional programs from prekindergarten through grade 12 should enable all students to—

Geometry

Analyze characteristics and properties of two- and three-dimensional shapes

Measurement

Apply appropriate techniques, tools, and formulas to determine measurements

Algebra

Represent and analyze mathematical situations and structures using algebraic symbols

Although much of the exploration of the solar system has focused on the planets closer to Earth, successful journeys have reached the outer planets. A series of Mariner and Pioneer spacecraft began humankind's initial reconnaissance of the solar system in the 1960s and 1970s. One spacecraft, Pioneer 11, became the first object from planet Earth to leave the solar system. Twin Voyager spacecraft mapped the four largest planets in the solar system. Their mission began with Jupiter in 1979 and concluded ten years later with Neptune. This time interval indicates the large distances that spacecraft must travel to explore our solar system.

Important Mathematical Ideas

Students create a model of the solar system in which both the sizes and the orbits of the planets are proportional to the same scale.

Mission

After their experiences developing a classroom model of the solar system, students focus on the outer planets. They have begun to realize that for a model of the distances to look reasonable, the outer planets must be far from the Sun. As an added dimension, this activity considers the size of the planets in the same model. Students use proportional reasoning to calculate both scaled distances between planets and scaled sizes of planets.

This activity culminates with an outdoor class session that requires careful planning to ensure the safety and adequate supervision of students and to select a proper place to conduct the activity.

Materials and Equipment

Calculators; string; Styrofoam balls or other spherical objects to represent the planets, such as circular poster-board or cardboard cutouts; and data sheet 5: "Planetary Diameters and Distances from the Sun," which can be found on page A-6 in the appendix.

Launching the Activity

In the activity "Our Solar System," students built a model of the solar system with scaled distances between planets, but they did not draw the planets to scale. In most depictions of the solar system, either the planets are not drawn to scale or the distances between planets are not drawn to scale, or both. In this activity, students attempt to create a solar system that is accurately scaled in both respects.

To initiate a discussion about the immensity of space, use the classroom model of the solar system and the class chart of data about the planets from earlier activities. The information gathered by students about the sizes of planets and their orbits may be useful. Ask the students to look at the size of Pluto in their model. "If we use the same scale for the size of Pluto that we used for the distance from Pluto to the Sun, how large would a model of Pluto need to be?"

Metric data are presented on data sheet 5. Customary measures are given in the margin of page 23.

In our classroom model, 40 feet represents 3,672 million miles, the mean distance of Pluto from the Sun. If we round 3,672 million miles to 4,000 million miles, then 40 feet represents 4,000,000,000 miles, or 1 foot represents 100,000,000 miles. One inch represents about 8,333,000 miles. Because the diameter of Pluto is not quite 1,500 miles, that planet should be represented by about 0.001 inch. Our model of Pluto will be very small—so small, in fact, that it is not practical to use in a model.

Discuss the difficulty of constructing a model that shows both the distances between planets and the sizes of planets on the same scale. Prompt students to think about a new scale for depicting the orbits of the planets by asking such questions as "How long is a football field?" and "Is our playground big enough to hold a football field?" Use this time to confirm students' conjectures. Going outside to measure 100 yards helps students as they think about longer distances.

Invite students to "pair share": "Discuss with a person near you the distances in our school neighborhood that are 200 yards long, 500 yards long, and 1000 yards long."

After the pair sharing, have students report their ideas to the class. Make a list of school-neighborhood distances that correspond to the distances being considered. At the conclusion of the sharing, tell students that the class is going to make a model of the solar system that is based on 1000 yards. The scaled distance from the Sun to Pluto will be 1000 yards.

Developing the Activity

Show students how to set up a proportion to determine the distances for other planets. A chart of mean distances allows the class to set up a ratio of the partial actual mean distance (Earth to the Sun) to the total actual mean distance (Pluto to the Sun).

By using miles, we can express the ratio as

$$\frac{93{,}000{,}000 \text{ mi}}{3{,}672{,}000{,}000 \text{ mi}}.$$

Using kilometers or astronomical units should give the same ratio. An astronomical unit, or AU, is the mean distance from Earth to the Sun. Because the distance from Earth to the Sun is 1 AU, we can determine the distance in astronomical units from Pluto to the Sun using the following equation:

$$\frac{93{,}000{,}000 \text{ miles}}{3{,}672{,}000{,}000 \text{ miles}} \approx \frac{1 \text{ AU}}{39.44 \text{ AU}}$$

NCTM Teaching Standards

The teacher [should convey] the notion that mathematics is a subject to be explored and created both individually and in collaboration with others.

(NCTM 1991, p. 115)

Mean Distances from the Sun

Mercury . . . . 0.39 AU, or 36 million miles

Venus . . . . 0.72 AU, or 67.2 million miles

Earth . . . . 1.00 AU, or 93 million miles

Mars . . . . 1.52 AU, or 141.5 million miles

Jupiter . . . . 5.20 AU, or 483.3 million miles

Saturn . . . . 9.54 AU, or 886.2 million miles

Uranus . . . . 19.21 AU, or 1,782.9 million miles

Neptune . . . . 30.06 AU, or 2,792.6 million miles

Pluto . . . . 39.44 AU, or 3,672 million miles

Magellan was deployed from Shuttle mission STS-30 in May 1989.

We can also use kilometers with the same result:

$$\frac{149,600,000 \text{ km}}{5,913,000,000 \text{ km}} \approx \frac{1 \text{ AU}}{39.44 \text{ AU}}$$

Each of these ratios is about 0.0253.

Some reference charts use slightly different values for these large distances; therefore, some differences might appear in the ratios. However, the ratios should be very close.

We want to express the same ratio for the scaled distances in the model. We know that the total distance from the Sun to Pluto is 1000 yards. The partial distance from the Sun to Earth (n) must be calculated and is shown as

$$\frac{n \text{ yards}}{1000 \text{ yards}}.$$

Using astronomical units, we can set up the two ratios as an equation:

$$\frac{n \text{ yd}}{1000 \text{ yd}} = 0.0253$$

$$n = 1000 \text{ yd} \times 0.0253$$

$$= 25.3 \text{ yd}$$

Because large numbers are involved in these equations, the use of technology is appropriate. Standard four-function calculators are sufficient, but fraction calculators or scientific calculators can be used.

Let students calculate the distances for each planet. The results should be entered into students' logbooks, shared with the whole class, and recorded on a classroom chart.

Teaching Tip

On a fraction calculator, once the ratio of the actual distances is set up, the student can use the fraction-to-decimal conversion key. The result is the decimal equivalent of the given fraction. This number represents the scaled distance of Earth from the Sun.

Next, similar procedures can be used to determine the scaled sizes for the planets. We know that the mean distance from Pluto to the Sun is 3,672,000,000 miles. This value can be converted to yards by multiplying the mean distance in miles by 1,760 yards/mile because a mile is 5,280 feet and a yard is 3 feet. Thus,

$$3,672,000,000 \text{ mi} \times 1,760 \text{ yd/mile}$$

$$= 6,462,720,000,000 \text{ yd}.$$

The scale is

$$\frac{1000 \text{ yd}}{6,462,720,000,000 \text{ yd}} = 0.0000000001547.$$

The diameters of the model planets should be in the same scale as the model distances. To begin this conversion, we start with the diameter of Earth, 7,921 miles.

The distance across the Earth, then, is 7,921 miles × 1,760 yd/mi, or 13,940,960 yd.

How many yards is that distance in the "1000-Yard Solar System"?

Let n = the diameter of the Earth model in yards.

$$\frac{n}{13{,}940{,}960} = \frac{1000}{6{,}462{,}720{,}000}$$

$$n = 0.0000000001547 \times 13{,}940{,}960 \text{ yd}$$
$$= 0.0022 \text{ yd}$$

Multiplying n by 36 converts the diameter from yards to inches, giving us 0.078 inch, which can be rounded to 0.1 inch. Earth, therefore, would be represented by a sphere approximately 0.1 inch in diameter.

At this time, students should use their calculators and the information in the class data chart to determine the scaled size of each remaining planet. These values should be recorded in the students' logbooks, shared with the class, and entered on the class chart. The illustration on page A-6 of the appendix might help students understand the relative sizes of the Sun and planets. Some time should be spent discussing real-world objects that are roughly equal to the scaled sizes of the planets.

Students are now ready for an important phase of the activity. They have calculated scaled values and should begin to think about these values as they apply to the school neighborhood. If the object representing the Sun were placed at the front of the school, where would the object for Earth be placed? Have students discuss the scaled distances with reference to the school's neighborhood and make estimates of where the planet objects would be placed.

In the culminating phase of this activity, have students mount the "planets" on poster board and clearly label them. Begin with the whole class at the Sun. Have one student hold the Sun while the rest of the class measures or paces off the scaled mean distances to the planets. To make the placement of the planets go more quickly, use the human pace as an approximation of a yard. Let students measure the length of the stride of several students to select one person whose stride is closest to 1 yard. Have this person pace off the distance to Mercury. Leave a student holding Mercury, and have the others proceed to place the remaining planets. In this way, students gain an understanding of the relative mean distances of the planets from the Sun and improve their intuitive notions of distance. The paced-off distances between proximate planets, as well as the scaled distances of the planets from the Sun, can be recorded by filling in a chart similar to the one on page A-7 of the appendix, which records sample student data.

Diameters of the Planets

Mercury . . . . 3,031 miles

Venus . . . . 7,514 miles

Earth . . . . 7,921 miles

Mars . . . . 4,215 miles

Jupiter . . . . 88,927 miles

Saturn . . . . 74,520 miles

Uranus . . . . 32,168 miles

Neptune . . . . . . 30,757 miles

Pluto . . . . ~1,447 miles

Concluding the Activity

To close the activity, ask students to write in their logbooks about their understanding of the size of our solar system. Ask them to comment on the mathematics they have learned while studying this activity.

For the debriefing, have students do three things:

- Ask them to complete their mission patches for this unit. They should try to incorporate important science ideas and mathematics concepts in the designs of the patches.
- The mission of this activity was to develop a model of the solar system. Ask students to write about the important features of a good model of the solar system and the mathematics that helped them make the model.
- The "Planets at a Glance" reference sheet in the appendix (A-6) can be given to students as a permanent record of some of their discoveries. Ask students to comment on what they learned about the solar system from this experience.

Mars *Global Surveyor* and *Pathfinder* patches

Beyond the Solar System

Looking at stars in the heavens allows us to see into the past. Photons from the visible spectrum of light emitted from distant stars travel millions and millions of miles to reach our eyes. The journey takes a long time. Thus, when we see a particular star, we are seeing the star where it was and how it looked when the light left the star. By seeing the star as it was, we are looking into the past.

A light-year is the distance that light travels in a year. At about 186,000 miles per second (about 300,000 km/sec), light travels approximately 5.8 trillion (5.8×10^{10}) miles in a year (about 9.46×10^{10} km/yr). In other words, the light that leaves a star that is 415 light-years from Earth will travel 415 years to reach our eyes. We are seeing the star where it was and how it looked 415 years ago. The constellation Pleiades, or the Seven Sisters, is approximately that distance from Earth. As we watch the evening skies for Pleiades, we are looking back 415 years in time.

The Eagle Nebula

As we look at the heavens with the naked eye, even in optimal conditions, we can see only about three thousand stars. Binoculars allow us to see more, and with a medium-sized telescope, we can see hundreds of thousands of stars. Large observatory telescopes open the vista to millions of stars. However, the same atmosphere that sustains life and protects us from radiation emanating from outer space also hinders our view of the heavens. For this reason, larger telescopes are often situated high on mountains. The remoteness ensures that no artificial light will reduce visibility, and the atmosphere is thinner at higher altitudes.

To make observations of the heavens with less interference from the atmosphere, scientists have used instruments in balloons,

The Coma Cluster

airplanes, sounding rockets, satellites, and the Shuttle orbiter. Longer-term and more powerful observations of distant objects in the universe were made possible by the launching of the Hubble Space Telescope (HST). This space-based telescope was launched from the Space Shuttle *Discovery* in April 1990. Flaws in the primary mirror of the HST limited early observations, but even the early images provided scientists with new data to study the universe. In 1993, NASA astronauts on the Hubble Servicing Mission made corrections to the mirror and other adjustments to improve the performance of the HST. More recently, during the fourth HST servicing mission (March 2002), spacewalking astronauts installed the Advanced Camera for Surveys (ACS). This camera allowed the collection of spectacular images from 420 million light-years away. We have looked deeper into space and, thus, farther back into time, thanks to the HST instrumentation.

The HST image shows three columns in the Eagle Nebula. A nebula is a cloud of interstellar gas and dust. The term *nebula* was coined to describe objects in space that appeared fuzzy when viewed through early ground-based telescopes. The Eagle Nebula is located in the constellation Serpens, the Serpent, alongside the southern Milky Way. It is approximately 7,000 light-years from Earth. Thus, the images we see of the Eagle Nebula show us how the nebula looked 7,000 years ago. This example illustrates how the HST allows us to look further into the history of the universe.

In a wider view, the HST image on the facing page shows a cluster of galaxies called the Coma Cluster. Although the Coma Cluster includes one thousand large and thousands of smaller galaxies, only hundreds of galaxies can be seen in this picture of a portion of the cluster. Some of the more easily identified ones are a large, bright, nearly circular galaxy; a spiral galaxy on the upper right side; and near the center, two colliding galaxies. This galaxy cluster is located 300 million (3×10^8) light-years from Earth.

The "Beyond the Solar System" module contains two related activities.

- "How Are New Planets Discovered?": In this introductory activity, students explore graphing from the perspective of the function being graphed. They use a motion detector to create graphs through motions made with their bodies. This activity prepares students to interpret graphs much like those analyzed by scientists in the discovery of new planets.

- "Interpreting Data from Newly Discovered Planets": Using knowledge about graph interpretation gained in the previous activity, students simulate the relative brightness of stars as they are transited by planets, then sketch graphs of what might be seen from the HST as scientists search for undiscovered planets.

ACTIVITY 1

How Are New Planets Discovered?

NCTM Standards

Instructional programs from prekindergarten through grade 12 should enable all students to—

Algebra

Understand patterns, relations, and functions

Use mathematical models to represent and understand quantitative relationships

Analyze change in various contexts

Measurement

Apply appropriate techniques, tools, and formulas to determine measurements

Astronomers use a mathematics skill called *graph interpretation* to help them discover new planets. The graphs of the light intensity of stars can show the existence of planets that cross, or *transit,* a star. The situation is similar to watching an insect fly in front of a large lightbulb, because stars are very large and planets are generally much smaller than stars. In this activity, the movement of students is used to create and interpret graphs.

Important Mathematical Ideas

The students will collect data, create graphs, and interpret graphs using calculator-based technology.

Mission

Scientists routinely interpret data from space that are received by land-based and space-based instruments. These data are often portrayed graphically for interpretation. To simulate this process in space science, students will interpret graphs of data collected using calculator-based laboratory instruments.

Materials and Equipment

At least one graphing calculator with a view screen and a motion detector, such as a Calculator-based Ranger (CBR), and a log sheet (see appendix, page A-2). Ideally, multiple setups should be available for small groups, but the activity can be completed successfully with one setup.

Launching the Activity

In preparation for this activity, a portion of the room will need to be cleared of all furnishings, such as tables and chairs. The clear zone should be a space with dimensions of 2 meters by 6 meters. The CBR should be placed at the center of one of the narrow ends of the clear zone and attached to the graphing calculator and view screen. Set the graphing calculator to use the Match program that comes with the CBR. This program allows students to walk in front of the motion detector to create a graph of distance over time. This experience will provide the background students need to interpret graphs of the change in light intensity over time in a way that is similar to that used by scientists to interpret data in their search for new planets.

Ask students to discuss the graph on the facing page with a partner. Have students share their questions about this graph or a similar one in which positive, negative, and zero slopes are represented but the scale and x- and y-axes are not labeled.

Students might ask what the graph represents, or they may want to know the values for the tick marks along the axes. They may be curious about the beginning and end of the graph or the way it changes as it goes up and down. Appropriate labels and vocabulary, such as *y-intercept, positive slope, negative slope, zero slope,* and *scale,* should be used whenever possible.

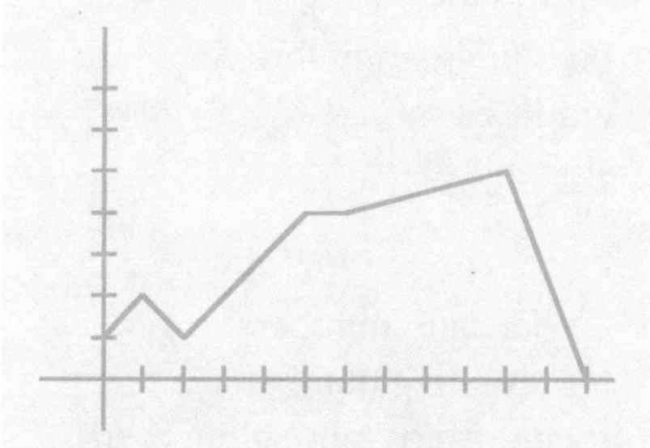

Developing the Activity

Once the students have exhausted their questions, run the Match program from the CBR. Choose a graph that displays distance in meters as a function of time in seconds. Tell students that this graph displays data that are similar to those displayed on the previous graph. Ask for a student volunteer to match the data on this graph by walking in front of the motion detector.

The first student to attempt matching the graph by walking will likely have some difficulty. Students tend to want to walk "across" the motion detector instead of walking forward and back in front of it. Instead of offering help yourself, ask the class to assist the student volunteer in a second attempt to match the same graph. Follow the directions on the screen to choose the same match for another try.

After the student has made two attempts to match the graph, ask for another volunteer to try to match a new graph. Repeat this process, allowing helpful suggestions from the class as long as the experience seems to be valuable.

Ask students what aspects of the graph helped them make a plan for matching the graph before starting data collection. If students talk about where they should stand to begin the graph, relate this point to the *y*-intercept. If students talk about how the graph helps them know when to move forward, move backward, and stand still, identify these situations as positive, negative, and zero slopes. The students may say that they use the tick marks along the axes to determine how far away to stand and for how long.

Ask students what each tick mark represents. They should respond that each tick mark along the *y*-axis represents a meter and that each tick mark along the *x*-axis represents a second.

Ask students how they might label the individual axes, as well as the entire graph. They should identify the *y*-axis as distance in meters and the *x*-axis as time in seconds.

After many students have had an opportunity to "walk" the graph, group the students in their mission teams to sketch a graph to correspond to the following story. Sketching diagrams of the attractions of the Kennedy Space Center that are mentioned in the story may be helpful. Tell the teams that they will share their graphs once they are complete. The graphs might be called *walking graphs* or something similar.

This year, I had my birthday party at Kennedy Space Center. When we arrived, my friends and I could see the Rocket Garden, the Astronaut Memorial, and the Space Shuttle. We were excited. The first thing we did was go to the ticket area. The line was long and moved very slowly. After we got our tickets, we walked through the

NCTM Standards

Instructional programs from prekindergarten through grade 12 should enable all students to—

Number and Operations

Understand numbers, ways of representing numbers, relationships among numbers, and number systems

Algebra

Understand patterns, relations, and functions

Use mathematical models to represent and understand quantitative relationships

Analyze change in various contexts

Measurement

Apply appropriate techniques, tools, and formulas to determine measurements

entrance building. Once outside, my friends and I ran to the Rocket Garden, where we played for a while. The Atlas rocket was the biggest of them all. My parents walked us around the pond to the Shuttle. We saw an alligator on the way. We climbed up the ramp to the Shuttle and went inside. Wow, is it small in there! After we were done, we ran down the ramp to meet my parents for lunch. We walked with my parents to the restaurant. Dessert was best. I had Space Dots!

As the teams share their graphs with the class, check to be sure that the graphs focus on how speed affects the slope of the graph.

As an additional check for understanding of graphs, give each team a graph of distance over time, then have the students work in their mission teams to create stories to describe the graphs. The following graph could be used for this purpose.

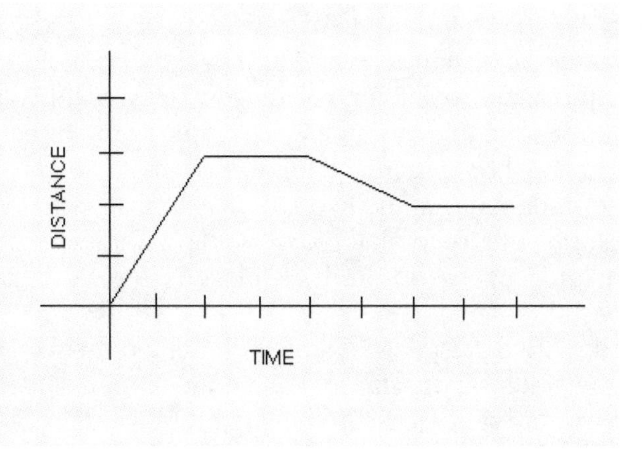

Extending the Activity

Using a map of Spaceport, have students make up a story that completes the birthday party from lunch until the people go back to the car. After the groups have written their stories, they can challenge their classmates to make graphs to match.

In their logbooks, have students reflect on the knowledge they used, what they learned, and what they would like to learn as a result of the experience from this activity.

ACTIVITY 2

Interpreting Data from Newly Discovered Planets

Using the Space Telescope Imaging Spectrograph aboard the HST, astronomers have succeeded in locating one planet that transits its star approximately every 3.5 days for 3 hours. What the astronomers really see is a graph that looks somewhat like the one here.

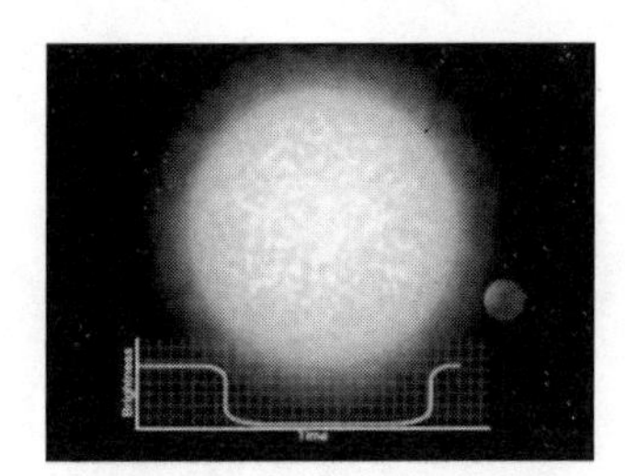

Important Mathematical Ideas

Students will use household objects related to light intensity to collect data, create graphs, and interpret graphs using calculator-based technology.

Mission

The use of space-based instruments reduces the distortion of our atmosphere. Images from HST provide much clearer data for astronomers to use in their graph interpretation. Students can use common objects to simulate the interpretation of light- intensity data and can search for Internet data to confirm their results.

Materials and Equipment

A flashlight, a paper tube with a diameter similar to that of the flashlight, a small piece of poster board, and at least one graphing calculator with a view screen and a motion detector, such as a CBR. Ideally, multiple setups should be available for small groups, but the activity can be completed successfully with one setup and a penlight for the extension.

Launching the Activity

Once students have experienced walking a graph and interpreting different situations that describe distance as a function of time, they should be prepared to investigate a functional relationship involved in the discovery of a new planet. Astronomers observed that the brightness of a particular star changed in a cyclical pattern. Using graphs of data from HST, astronomers concluded that the star appeared less bright when a planet was "in front of" the star during the planet's orbit.

Internet Resource

Animations of planets transiting stars can be found by searching for "planet transit" at the following Web site:

hubblesite.org/newscenter/newsdesk/archive/releases/2001/38/video/c

To understand this notion and to be successful with this activity, students need to think about relative brightness.

Ask students to share examples of situations in which things seem to change in brightness, such as—

- that the moon seems to shine more brightly on a clear night than an overcast one;
- that a light bulb's glow is dimmed by the lampshade;
- that car headlights can be dimmed with a dimmer switch when approaching an oncoming car; or
- that a flashlight beam is greatly dimmed when you put your hand over the light.

Developing the Activity

Show students the flashlight example, then either insert the flashlight into a paper tube or roll a piece of paper around the flashlight. Darken the room. Shine the flashlight at a blank wall, and begin the discussion of brightness.

The light on the wall represents 100 percent of the light. Next cover half of the flashlight lens with a small piece of poster board. Ask what brightness, in percent, is now represented. Students should agree that the brightness is approximately 50 percent of the original brightness. They will determine this percent from looking at the flashlight, not the light on the wall.

Slowly pass the poster board across the flashlight until the light is totally obscured, then keep moving the poster board until the full brightness is restored. Repeat this process, stopping at three or four points to ask what percent of the light is shining on the wall.

Have students work in pairs to sketch graphs that represent the *intensity* (relative brightness) of the light from the flashlight over the time required to move from full brightness to complete darkness to full brightness again. Students may need several tries with the flashlight to sketch a graph.

The graphs might look something like the one shown below.

Have students share their graphs and their justifications for the sketches they made.

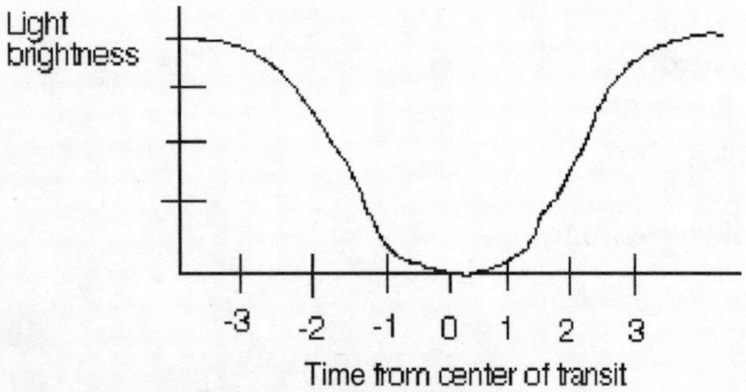

Relate this activity to the work of astronomers. Tell students that the flashlight represents a star. Glue a small piece of poster board (about the size of a grape) to a wire, and hold the object in front of the flashlight. This object represents a planet.

With the help of a student, simulate the orbit of this "planet" (the small piece of poster board) around the "star" (flashlight). Students, working in pairs, can sketch a new graph of the light intensity when the planet transits the star.

Concluding the Activity

Discuss the sketched graphs in class, and have students compare their results with those of astronomers by looking for NASA Web sites giving new planet data.

Interestingly, astronomers have interpreted the reduction in light to be caused by the planet's blocking some of the light from the star. With the refined instruments on HST, the planet's size, mass, and orbit can be detected. Graph interpretation is an important skill for astronomers who are trying to discover new planets.

Extending the Activity

Ask students to think about a small star orbiting a larger star, and have them work with partners to sketch a graph of a complete orbit of this situation.

To help students think about one star orbiting another, use the same flashlight and a penlight inside a paper tube. The penlight represents the smaller and dimmer star. During the simulation, the light of the penlight should point away from the flashlight at all times in the orbit.

Ask students how the graph would be affected if the light from the smaller star was very intense, that is, as bright as, or brighter than, that from the larger star. Also, the penlight shines in only one direction, whereas a star would shine in all directions during the orbit. How does this factor affect the graph?

Students might enjoy representing this mission of planet discovery with a mission patch.

Space and Speed

"I think I can, I think I can, ..." These familiar words from *The Little Engine That Could* might very well be the sentiments of *Voyager 1* and *Voyager 2.* These two remarkable spacecraft were launched in 1977, and they are still traveling through space today. As of January 2001, *Voyager 1* was at a distance of 12 billion kilometers (80 AU) from the Sun and *Voyager 2* was at a distance of 9.4 billion kilometers (62.7 AU) from the Sun. On the twenty-fifth anniversary of their respective launches, *Voyager 1* is approximately 7.8 billion miles from Earth and *Voyager 2,* which was launched first, is about 6.3 billion miles away.

The original destinations of the Voyager spacecraft were Saturn and Jupiter, but their 4-year missions were extended, and they have traveled far beyond these planets. In 1989, *Voyager 2* made a close flyby of Neptune, the final outer planet in our solar system visited by a Voyager spacecraft. *Voyager 1* completed its planned close flybys of the Jupiter and Saturn planetary systems while *Voyager 2,* in addition to its own close flybys of Jupiter and Saturn, completed close flybys of the remaining two gas giants, Uranus and Neptune. *Voyager 1* is escaping the solar system at a speed of about 3.6 AU per year, and *Voyager 2* is leaving at a speed of about 3.3 AU per year.

Both Voyager spacecraft are headed toward the outer boundary of the solar system in search of the *heliopause,* the region where the Sun's influence wanes and the beginning of interstellar space can be sensed. No spacecraft has ever reached the heliopause. The Voyagers may be the first to pass through this region, which is thought to exist at some point 5 to 14 billion miles from the Sun. Sometime in the next 10 years, the two spacecraft should cross an area known as the *termination shock.* In this region of space, the million-mile-per-hour solar winds slow to about 250,000 miles per hour, giving the first indication that the wind is nearing the heliopause. The Voyagers should cross the heliopause 10 to 20 years after reaching the termination shock.

The Voyagers have enough electrical power and thruster fuel to operate at least until 2020. By that time, *Voyager 1* will be 12.4 billion miles from the Sun and *Voyager 2* will be 10.5 billion miles away. Eventually, the Voyagers will pass other stars. In about 40,000 years,

Voyager 1 will drift within 1.6 light-years (9.3 trillion miles) of AC+79 3888, a star in the constellation of Camelopardalis. In some 296,000 years, *Voyager 2* will pass Sirius, the brightest star in our sky, at a distance of about 4.3 light-years (25 trillion miles). The Voyager spacecraft are destined—perhaps eternally—to wander the Milky Way.

Distances and the speed of travel in space are difficult to conceptualize. This chapter is intended to give students some ways to think about the vastness of space and the speeds associated with various forms of travel.

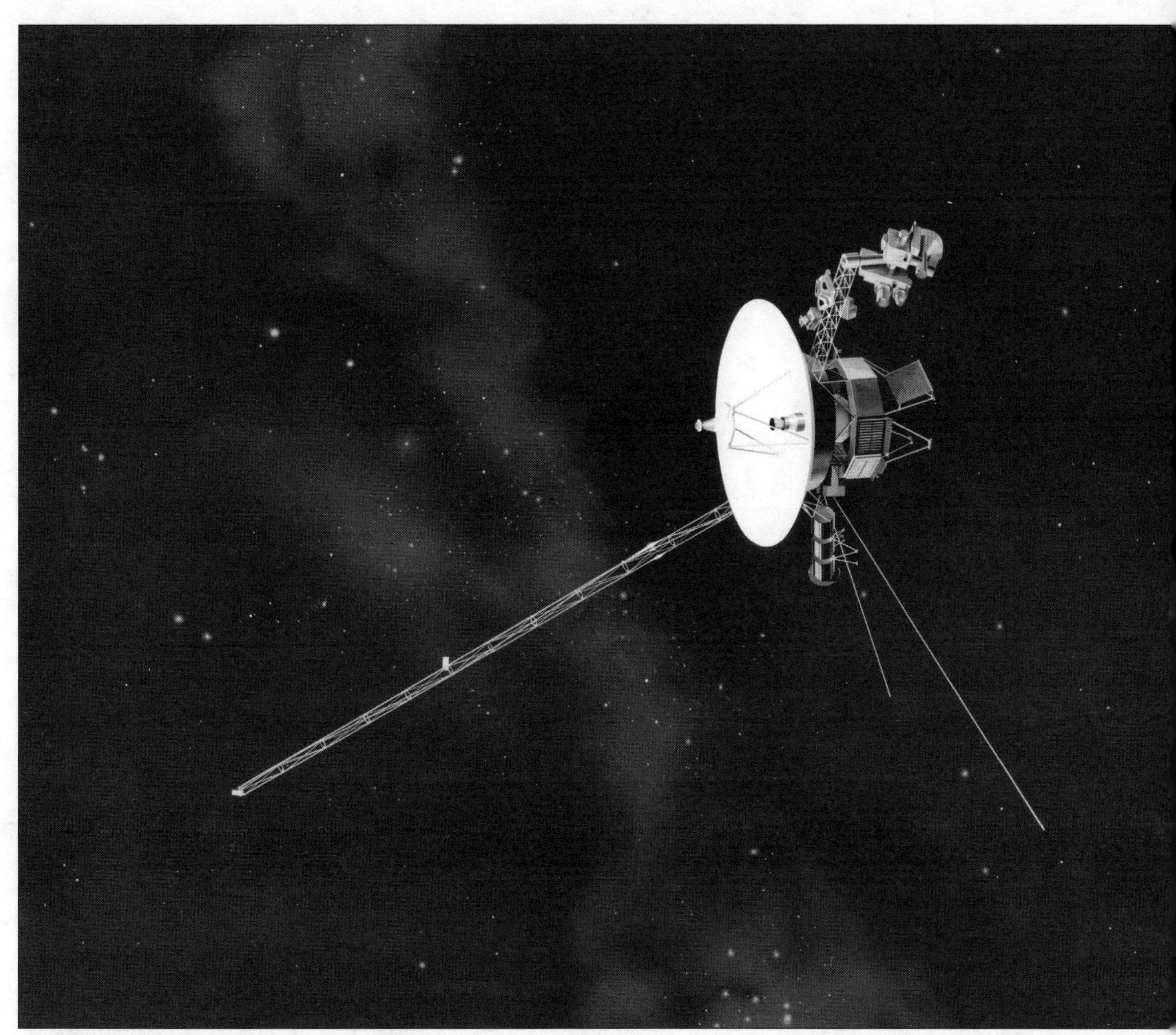

Artist's conception of Voyager

This is the large galaxy in the constellation Andromeda. The Andromeda galaxy is the nearest large galaxy to our own. The image is a mosaic of ten separate Galaxy Evolution Explorer images taken in September, 2003. The star forming arms are unusual in being quite circular rather than the usual spiral shape. Several companion galaxies can also be seen. These include Messier 32, a dwarf elliptical galaxy directly below the central bulge and just outside the spiral arms, and Messier 110, which is above and to the right of the center. NASA Jet Propulsion Laboratory (NASA-JPL)

Stars Afar

Space is vast, and the distances of space are difficult to comprehend. The facts listed below are examples of the enormity of space.

- Light travels at about 186,000 miles per second, or about 300,000 kilometers per second.
- A light-year is a measure of the distance that light travels in a year.
- Sirius, a star fairly close to Earth, is 8.6 light-years away.
- Even a bright star, such as Alnair, is more than 100 light-years away.
- The Milky Way, our galaxy, is about 100,000 light-years in diameter.
- The most distant things that astronomers can see are about 18,000,000 000 light-years away.

In the "Stars Afar" module, students will explore the meaning of large numbers as they are used to describe some aspects of the distances in space. This module contains the following two activities:

- "Speed of Light": Students become more comfortable with large numbers as they work through calculations to determine the distance light travels in a year.
- "What Light Are We Seeing?": Students come to understand the vast distances between Earth and even the closest stars by calculating how long ago light left the stars for us to see it today.

ACTIVITY 1

Speed of Light

NCTM Standards

Instructional programs from prekindergarten through grade 12 should enable all students to—

Number and Operations

Understand numbers, ways of representing numbers, and relationships among numbers

Measurement

Understand measurable attributes of objects and the units, systems, and processes of measurement

Teacher Note

Be certain to point out to students that the Casio calculator from for scientific notation,

1.4⁸ for 140,000,000,

should not be confused with 1.4 to the 8th power.

Important Mathematical Ideas

The light-year is the standard measure used to describe the vast distances in space. Students extend their understanding of a light-year as a measure of distance.

Mission

Students explore large numbers and calculators using the motivating topic of distances to stars.

Materials and Equipment

Each student needs a four-function calculator.

Conducting the Activity

The speed of light is approximately 186,000 miles per second. Note that this speed is not measured as miles per hour but miles per *second.* Have students use calculators to answer the following questions, then read and discuss their results.

How many miles does light travel in a minute?

$$186{,}000 \text{ mi} \times 60 \text{ sec} = 11{,}160{,}000 \text{ mi/min}$$

How many miles does light travel in an hour?

$$11{,}160{,}000 \text{ mi} \times 60 \text{ min} = 669{,}600{,}000 \text{ mi/h}$$

As a comparison, the Space Shuttle orbits the Earth at approximately 17,000 miles per hour. How does the speed of the Space Shuttle as it orbits compare to the speed of light? To answer this question, have students perform the following calculation:

How many miles does light travel in a day?

$$669{,}600{,}000 \text{ mi} \times 24 \text{ h} =$$

Point out that some calculators show a different type of number on the display. Because the correct answer has too many digits for the display, different "forms" of scientific notation are used to express the result.

$$1.60704^{10} \text{ or } 1.60704 \times 10^{10}$$

Multiplying 1.60704 by 10^{10} effectively moves the decimal point ten places to the right.

$$1.60704 \times 10^{10} = 16{,}070{,}400{,}000 \text{ mi/day}$$

Have students read and discuss this large number.

To determine the speed of light in a year, multiply the speed of light in a day by 365 days in a year.

$$1.60704 \times 10^{10} \text{ mi} \times 365 \text{ days} =$$

Once again, the calculator shows its result using scientific notation.

$$1.60704 \times 10^{10} \text{ mi} \times 365 \text{ days} = 5.865696^{12}$$

Remind students that 5.865696^{12} represents

$$5.865696 \times 10^{12}, \text{ or } 5{,}865{,}696{,}000{,}000.$$

Have students read and discuss this large number.

Students now know that light travels

5,865,696,000,000

miles in one year. This distance is called a *light-year.*

Concluding the Activity

This activity can be done using the metric system if preferred, or the metric equivalents can be explored as an extension of the activity. Light travels at 299,792,458 meters per second. This fact can be used to derive the speed in kilometers per second and, ultimately, the distance light travels in a year using metric measures.

Teaching Tip

Although the early calculations in this activity can be done by hand, students should still use calculators. The goal of this activity is for students to read and understand large numbers, not to practice calculation skills.

Directions for a Casio fx 65

PRESS shift sci not

ENTER 5865696 =

The display will read 5.865695^{06}.

ENTER × 1000000 =

The display will read 5.865695^{12}.

PRESS M +

ENTER 25222 =

The display will read 2.5222^{04}.

ENTER × 1000000000 =

The display will read 2.5222^{13}.

PRESS MR ÷ =

The display will read 4.2999.

Directions for a Texas Instruments TI-34II

ENTER 25222000000000 =

The display will read 2.5222×10^{13}.

ENTER ÷

PRESS 5865696000000 =

The display will read 4.299915986.

ACTIVITY 2

What Light Are We Seeing?

NCTM Standards

Instructional programs from prekindergarten through grade 12 should enable all students to—

Number and Operations

Understand numbers, ways of representing numbers, and relationships among numbers

Algebra

Understand patterns, relations, and functions

Measurement

Understand measurable attributes of objects and the units, systems, and processes of measurement

Important Mathematical Ideas

Students use their abilities to read large numbers, apply calculator skills to complex operations, and use scientific notation to represent large numbers. Further, students gain a deeper understanding of large numbers by reflecting on the distances of selected stars from Earth.

Mission

Students compare a time line of historical events to the time required for light to travel from selected stars to Earth.

Materials and Equipment

Each student needs a four-function calculator with a memory key and scientific notation features.

Launching the Activity

Even though light travels at a great rate of speed, for it to travel large distances takes time. Because stars are vast distances from Earth, their light takes time to reach us. This activity explores how much time is needed for the light from a star to reach Earth.

Our nearest star, other than the Sun, is approximately 25,222,000,000,000 miles away from us.

Have students read and discuss this large number. Remind them that this star is our nearest neighbor star.

Because the distances to stars are so great, scientists decided to use the distance that light travels in one year, a light-year, as one way to describe distances to stars.

Developing the Activity

Thinking about the distance to our nearest star neighbor in light-years requires some calculations.

We know the distance to the nearest star in miles. If we divide this distance by the number of miles light travels in a year, we will find the number of years the light travels to reach us.

$$25{,}222{,}000{,}000{,}000 \text{ mi} \div 5{,}865{,}696{,}000{,}000 \text{ mi/yr}$$

Students will need to use scientific notation on the calculator to do these calculations.

Dividing these two numbers on the calculator tells us that light from the nearest star requires approximately 4.3 years to reach us. Space scientists use the distance that light travels in a year as a unit for measuring large distances in space. So, they would say that Proxima Centauri is 4.3 light-years from Earth.

On page A-6 in the appendix is a chart that shows the distances in light-years (ly) of some of the stars closest to us.

Star	Distance (ly)
Proxima Centauri	4.3
Alpha Centauri	4.4
Barnard's Star	5.9
Sirius	8.6
Ross 154	9.7
Ross 128	10.9
Procyon	11.4
Luyten's Star	12.4
Altair	16.8
Fomalhaut	25.1
Vega	25.3
Pollux	34
Arcturus	37
Capella	42
Castor	52
Aldebaran	65
Regulus	78
Alioth	81
Menkalinan	82
Alkaid	101
Alhena	105
Duhbe	124
Kaus Australis	145
Alphard	177
Peacock	183

Concluding the Activity

Prompt students to think about the star chart (p. A-6) by asking the following questions:

- How many miles away is the star Ross 128?
- Light from Sirius takes how many times longer to reach Earth than light from Proxima Centauri?
- From what star does light take nearly twice as long to reach Earth as that from Barnard's Star?

Continue the discussion by asking students to think carefully about the following topics:

A light-year is the distance that light travels in a year. If you see light that traveled 1 light-year, the light started from its origin 1 year ago. In the instance of our nearest star, Proxima Centauri, the light we see started traveling toward Earth 4.3 years ago. The light that we see from some stars started traveling at about the same time that you were born.

Choose the star in the chart that is about the same distance from Earth in light-years as your age in years. Light from that star has been traveling all your life to reach your eyes here on Earth.

Make a list of the stars from which light has been traveling less time than your age in years to reach Earth. Do you think more stars are at distances (light-years) greater than your age or less than your age?

Extending the Activity

Light from many stars reaches Earth each day, and the stars themselves are varying distances from Earth. Thus, the light arriving from different stars started at different times. For example, Vega, a bright star in the night sky, is 25.2 light-years away. Light arriving on Earth today from Vega started traveling 25.2 years ago.

Ask students to determine in what year and month the light arriving from Vega today began traveling to Earth.

Remind students that 25.2 years is not the same as 25 years and 2 months. Students should use the calculator to convert 0.2 years to months.

$$\frac{2}{10} = \frac{x}{12}$$

Students may even convert the 0.2 years to days to be more accurate.

Find out who in the class has a relative with a birthday in the month and year that light left Vega.

At the same time that light arrives on Earth from Vega, light also arrives on Earth from the star Pollux. This star is 34 light-years from Earth, meaning that the light from Pollux took 34 years to reach Earth. What year did it begin its journey? Ask students to think of a special event from history that happened in that year.

Have students make a time line to show when starlight left certain stars and try to tell one special thing from history that happened in that year.

The time line below shows today. Have students write the year under the far right tick mark on the time line. Because Pollux is 34 light-years away, students must subtract 34 from the current year to find out when light arriving on Earth this year left Pollux. Allow students to talk with partners or group members to think of important things that happened 34 years ago.

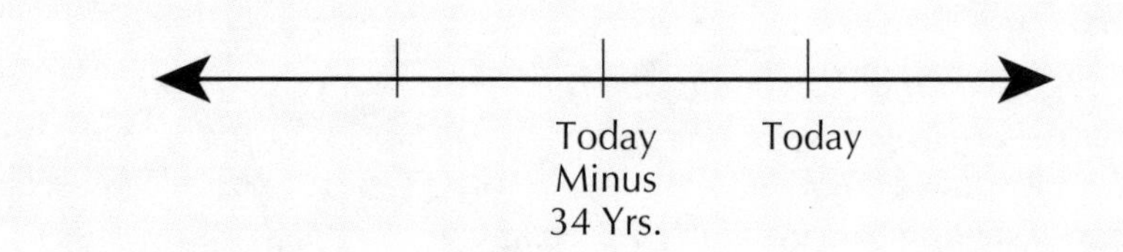

On page 41 is a table of stars and their respective distances from Earth. Ask students to think about the distances from Earth of these stars and the stars shown in the first chart. Can students name stars from which light left before their parents were born? Think of another important date in history. For example, when did an astronaut first walk on the Moon? Can students name a star that is about that distance from Earth?

How Far Can I Go in 8 Seconds?

Space travel is an idea that intrigues young and old alike. Can we go to other planets? Can we travel at "warp speed"? Thinking about space travel allows the imagination to run wild. What are some of the difficulties of space travel? In this module, students use important mathematics to develop an understanding of long distances, time, and average rate of speed. Some of the activities in this module have links with the activities in "Learning about the Solar System" and "Beyond the Solar System." The activities in these modules can be intermingled if students' interest in the topic leads in that direction.

The three activities presented can span from six to nine class periods. These activities need not be conducted on sequential days. One or two activities a week can be completed as time allows. Some portions of the activities are optional. Likewise, extensions for the activities can be done according to students' interests and abilities. Options for assessment activities are also included.

- "Finding Our Top Speed": Students determine the length of time required for a student to walk a given distance. After collecting several types of data, students use graphs to think about the meaning of the data. In one instance, they enter data on a graph of distance at a constant time. This graph allows students to think about slope as a representation of speed. The "steeper" the slope is, the faster a student has walked. This activity reinforces the graphing exercise in "How Are New Planets Discovered?"

- "An 8-Second Trip": In this activity, students explore the concept of speed using familiar objects (toys) and apply this new knowledge to well-known means of transportation, including the Space Shuttle. This activity has a great deal of flexibility.

  Three experiences with toys are presented. Ideally, all three activities should be conducted; however, if a lack of time or materials poses a problem, students will profit from any of the hands-on experiments that can be done. Experiences with the speed of toys make the concept of speed applied to real modes of transportation more meaningful.

- "Travel in the Solar System": This activity focuses on human travel in space. The distances from one planet to another are considered in relation to the time

A Space Shuttle blasting off

required for humans to reach the planets for exploration. Students use estimation, computation, and algebraic reasoning as they consider the time required to travel to a planet and to return to Earth. This activity has obvious links with the "Learning about the Solar System" module. Students may want to "loop back" into the earlier unit to refine their thinking. This backtracking should be encouraged if students' interest in space is high. Teams and individuals should be encouraged to reflect on their learning in earlier activities.

The activity "An 8-Second Trip" requires advance preparations and approval from the school administration. Materials for the activity must be sought long before the module is initiated, and planning for the use of the materials is necessary to ensure success. Establishing stations for the materials required for each activity is one strategy to allow all three activities to be performed if materials are limited.

Students formulate their thoughts and write about their learning throughout the module. The logbook form of journal writing allows students to express their understanding of mathematics, construct the meaning of previously encountered concepts, and internalize learning. The logbooks are vehicles for students to record their mathematical understanding, as well as their understanding of space science.

Creating mission patches and logbook covers are important parts of the learning in this unit. Some students can express geometrically their understanding of mathematics applied to space science. The design and production of mission patches and logbook covers give students who use spatial sense extensively an avenue to communicate and internalize learning. In making the patches and logbook covers, students should be encouraged to portray how mathematics is important in the activities in the unit.

ACTIVITY 1

Finding Our Top Speed

"Are we there yet?" "How much farther?" These questions are commonly asked by young people on long automobile trips, indicating their poor concept of the passage of time and of distance related to time. This lack of understanding carries over into students' concepts of time and travel in space.

In December 1995, the probe that had been released by the *Galileo* spacecraft in July 1995 entered Jupiter's atmosphere. *Galileo* had traveled 2.3 billion miles since its launch in October 1989. It spent the first 3 years in the inner solar system. During one flyby of Venus and two flybys of Earth, it gathered enough velocity from the gravity of the planets to reach Jupiter. Throughout its long journey, *Galileo* has been sending data about the solar system back to Earth.

By studying the passing of time and time versus distance, students refine their thinking about the time required to travel the long distances in space. Of course, the long time required for this type of travel is exactly why it is difficult for humans to make journeys to the planets. The question "Are we there yet?" has real meaning for space travel. Remember, the *Galileo* spacecraft traveled for more than 6 years one way to get to Jupiter.

This activity sets the stage for a discussion of travel in the solar system. By exploring an Earth-bound, real-world, hands-on activity, students develop an understanding of time and distance. The mathematics necessary for the activity focuses on measuring time and distance and graphing to present and interpret the data collected.

Important Mathematical Ideas

Students use computation, estimation, and algebraic reasoning to develop their understanding of the problems posed by the vast distances involved in space travel.

Mission

Students explore the potential of human travel in space through hands-on activities that develop their understanding of average rate of speed. They determine the length of time needed to walk or run a given distance, then plot the data on a graph. They also determine the distance a subject can walk in 8 seconds, plot the data on a graph, and use the results to develop an understanding of the concept of slope.

NCTM Standards

Instructional programs from prekindergarten through grade 12 should enable all students to—

Number and Operations

Understand meanings of operations and how they relate to one another

Algebra

Understand patterns, relations, and functions

Analyze change in various contexts

Measurement

Apply appropriate techniques, tools, and formulas to determine measurements

Data Analysis

Develop and evaluate inferences and predictions that are based on data

NCTM Teaching Standards

The teacher of mathematics ... fosters the development of each student's mathematical power by providing and structuring the [class] time necessary to explore sound mathematics and grapple with significant ideas and problems.

(NCTM 1991, p. 57)

The teacher of mathematics should orchestrate discourse by—

posing questions and tasks that elicit, engage, and challenge each student's thinking ... [and]

deciding when to provide information, when to clarify an issue, when to model, when to lead, and when to let a student struggle with a difficulty.

(NCTM 1991, p. 35)

Materials and Equipment

Stopwatches, yardsticks or 50-foot measuring tapes, masking tape, index cards, and graph paper

Launching the Activity

Start the activity by asking students how far they can go in 8 seconds. Typically, students answer in a variety of ways. Some respond with distances ranging from a few feet to the length of a football field. Others ask, "Are we traveling by foot, on a bicycle, or in a car?" Confine the discussion at this point to travel on foot.

The passing of time is a difficult concept for everyone. In certain settings, when we are enjoying ourselves, time seems to fly by; in other contexts, time seems to stand still. To help students develop their sense of time in a neutral context, ask them to close their eyes. Tell students when to start, and ask them to raise their hands after exactly 1 minute has passed. Practice with estimating the passing of time improves performance. Ask students to share the techniques they used to guess about the length of a minute. Let them practice with time intervals of less than a minute. Finally, have students estimate the duration of 8 seconds.

After students have practiced estimating the passing of time, they are ready to see how far a teacher can walk in 8 seconds. Give one student a stopwatch, and have him or her time a teacher walking from the front to the rear of the classroom. The teacher's rate of walking might be a topic of discussion. Students often feel that they can walk farther than a teacher because they can walk faster.

Developing the Activity

To gather data about students' walking speeds, mark off, in a school hallway or outdoors, distances from 25 feet to approximately 100 feet in increments of 5 feet. Mark the intervals with masking tape. Have each student carry an index card that is labeled with his or her name and has lines for recording as many trials as will be completed.

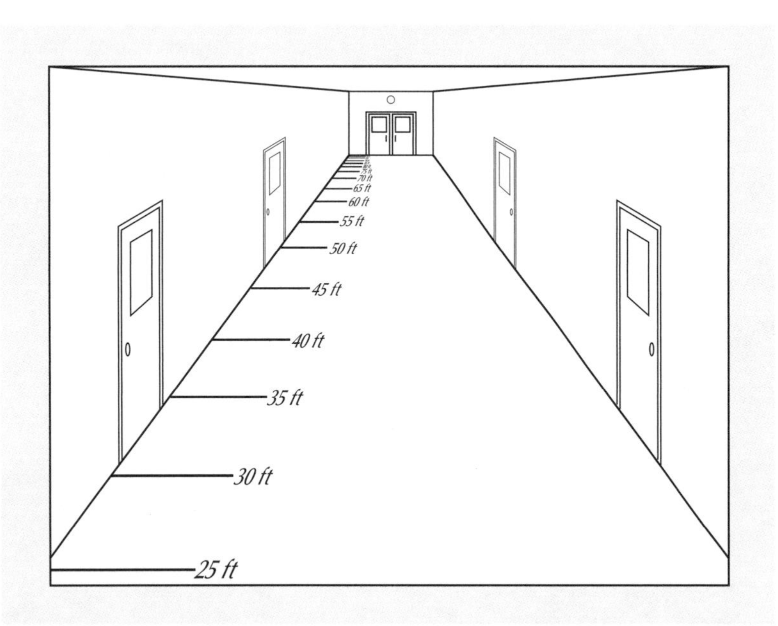

A school hallway marked off in 5-foot increments

For this first experiment, everyone walks 100 feet, and the timers tell students the time each took to complete the distance. Each mission team takes a turn timing another team.

Line up students, and begin the trials. Tell members of a mission team when to go. Three students should have stopwatches to time each walker. The time recorded is the median of the three times shown on the timers' watches. Working with multiple timers avoids losing data because of difficulties using a stopwatch. If enough stopwatches are not available for this approach, have each timer pick a participant and keep time for that student. Some students may need to repeat their walks if timers make mistakes.

To facilitate making a graph of the class data, have each student record his or her time on the reverse side of the index card in large writing. Begin by making a graph of the class data using the students themselves. Ask five students to come to the front of the room and stand in order according to their times, from the least to the greatest. Ask another group to come to the front and put themselves into the ordered group. If the times are the same, the students should stand behind one another. When all students are in the ordered group, the graph is complete.

Record on the chalkboard or an overhead transparency a frequency table for the human graph. When seated again, each student should make a bar graph of the data in the frequency table.

Note that this part of the activity may require an entire class period.

The next phase of this activity requires students to collect data about how far they can walk in 8 seconds. That is, the time allowed for walking is held constant and the distance varies from student to student. The marked-off increments of 5 feet are used to measure the distances.

Only one timer is necessary. A teacher is probably the best timer for this activity because the stop-and-go commands need to be authoritative.

At the stop command, each student looks at the distance markers and records the distance walked. Of course, students do not always stop on a mark. They need to agree on how to estimate the number of feet they have walked beyond a mark, add that distance to the marked distance, and record their results on their index cards.

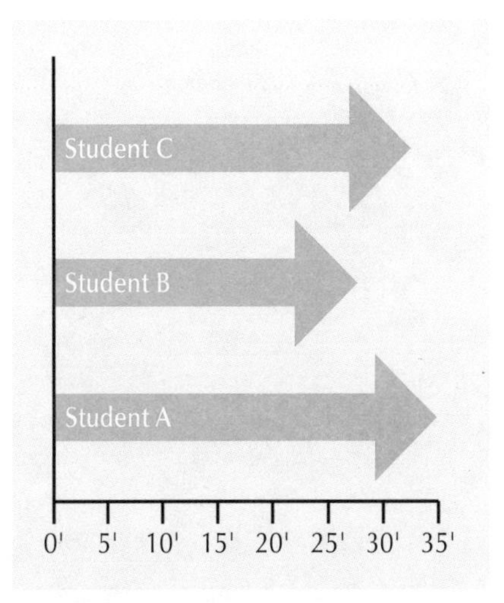

Different students can walk varying distances in 8 seconds

Concluding the Activity

One trial is sufficient for this activity, but multiple trials can be done. For multiple trials, use an odd number of trials and the median of the results for the final graphing activities.

Have each mission team record its results on one graph. The graph should show distance versus time, with distance on the vertical axis and time on the horizontal axis. Be sure to talk about the point where time is zero and distance is zero, which is also a potential data point. On the graph, connect each point (representing how far each student on the mission team walked in 8 seconds) to (0, 0). The steepness of the lines connecting (0, 0) with the data points shows the average rate of speed, or slope. Use this opportunity to discuss the concept of slope and to give a formal definition of the slope of a line.

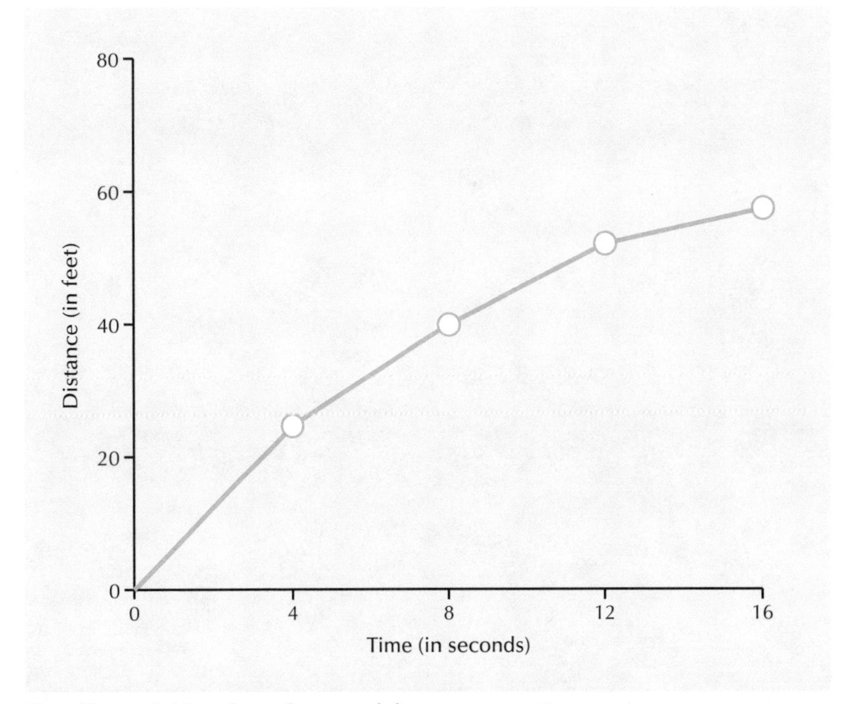

The effects of tiring show that speed decreases over time

Two types of graphs and various statistical concepts are used in the activity. To close the activity, students should use their mission logbooks to record their thinking about these ideas.

Students can also make mission patches for "Finding Our Top Speed." As time allows, students begin working on visual interpretations of what they are learning. Listing ideas to incorporate into the patch design is a good practice. The list should include both mathematical and scientific ideas.

Extending the Activity

Students may want to check their speeds over longer distances to confirm that they go twice as far in 16 seconds. Note that the empirical data most likely do not show a linear relationship. As the length of time increases, the average speed usually decreases. Talk about the effects of tiring over time. If shorter times are used, the speed could be faster. Point out that the linear graph approximates where each walker would be in the intervening times if he or she walks at a constant rate of speed. This result explains why we characterize the graph as showing the average rate of speed.

ACTIVITY 2

An 8-Second Trip

With car chases and thrilling airplane maneuvers in action-based movies, students have witnessed many events involving speed. In spite of this exposure, students have little experience trying to quantify speed and compare rates of speed. This activity involves several experiments using familiar objects from the students' world with which to explore the concept of speed. These hands-on experiences are used as bases for increasing students' understanding of the speed of conventional modes of transportation.

The goal of teaching mathematics is to help students gain mathematical power. They need experiences using mathematical reasoning, solving problems, and making connections with the real world. In this activity, students explore the concept of speed as it relates to travel in space. The mathematics of this topic is rich. Students use algebra, measurement, geometry, and statistics to improve their understanding of speed.

NCTM Standards

Instructional programs from prekindergarten through grade 12 should enable all students to—

Algebra

Analyze change in various contexts

Measurement

Apply appropriate techniques, tools, and formulas to determine measurements

Important Mathematical Ideas

Each of the experiences in this activity is designed to give students first-hand knowledge about average rates of speed. Rates of change are difficult for students to comprehend; thus, multiple experiences are important.

Mission

Students collect, display, and analyze data about distance, rate, and time to gain an appreciation of the speeds involved in space travel. Students determine the speeds of various toys by conducting experiments, then calculate the speeds of various methods of transportation by using algebra.

Materials and Equipment

A log sheet (see appendix, page A-2) and the following materials for each experiment:

Experiment using toy cars

At least one toy car, at least 3 feet of track for a ramp, a board or strong yardstick, another yardstick or tape measure, masking tape, and stopwatches

Experiment using battery-powered remote-control cars

One or more cars, stopwatches, yardsticks or measuring tapes,

The Speed Demon rocket car can attain speeds of up to 300 mi/h. (Photo courtesy of Brent McNeeley, Hell on Wheels, Inc.)

masking tape, "traffic" cones from the physical education teacher, and extra batteries

Experiment using toy rockets

At least three rockets for each class, one rocket launch pad, a 12-volt battery, and stopwatches. Clinometers or hypsometers would be necessary to make indirect measurements using basic trigonometry.

Launching the Activity

In the "Finding Our Top Speed" activity of this module, students learned how far they could move in 8 seconds. During the opening discussion, some students may have talked about traveling at high speeds in cars and airplanes. From students' suggestions, list ways to travel fast. This list will probably contain several types of automobiles and airplanes, as well as other vehicles. Consolidate the list by considering only broad categories, such as bicycles, motorcycles, cars, airplanes, and jet airplanes. The Space Shuttle can be a category in itself.

Continue the discussion by asking students to recommend some miniature test vehicles, given that the speed of jet airplanes and cars cannot be tested in the classroom or around the school.

Usually, every class includes a few students who collect small toy cars that come with ramps and loops. Other students may own battery-powered remote-control cars. Some may have experimented with small toy rockets. Some or all of these vehicles can be used in and around school to help students collect data about speed.

This activity is most effective if students have an opportunity to collect and organize data about the speed of familiar objects. This experience is valuable before they consider automobiles, airplanes, and other means of transportation in the adult world.

Specific directions follow for experiments with three sets of familiar objects: toy cars, remote-control cars, and small toy rockets. Provide as many of these experiences as possible.

Conducting the Experiment Using Toy Cars

Toy cars rolling down a ramp may not remain in motion for 8 seconds. Students need to time the cars and calculate the speed of the cars using the formula $d = rt$, in which d = the distance traveled, r = the average speed of the car, and t = the time traveled. The average rate of speed (r) can be calculated by manipulating the time-distance-rate formula algebraically.

Toy cars on a ramp

NCTM Teaching Standards:

Teachers of mathematics should pose questions that are based on—

sound and significant mathematics;

knowledge of students' understandings, interests, and experiences;

knowledge of the range of ways that students learn mathematics.

(NCTM 1991, p. 25)

The teacher of mathematics should promote classroom discourse in which students—

... use a variety of tools to reason, make connections, solve problems, and communicate.

(NCTM 1991, p. 45)

Assemble approximately 3 feet of track. To keep the track straight, secure it to a board or strong yardstick. Elevate one end, and roll a toy car down the track.

Although this activity has "8 seconds" in the title, the stopping time for a toy car is not likely to be exactly 8 seconds. The ramp's surface and other variables can affect the duration of a "run." Before assigning students to working groups, allow them to make a few trial runs to determine a reasonable time for the car to come to a stop. All groups will use the same time interval, for example, 5 seconds. This interval becomes a constant in the data.

Group the class into mission teams of four students to collect data. One student is the car starter, another is the recorder, and two are timers. Students can rotate tasks for different trials and ramps.

Have each mission team describe its ramp using the ratio of the height of the ramp to the horizontal length of the ramp. Have students collect three to five sets of data about the car and any given ramp. They will measure the distance traveled over the agreed-on constant time.

Students may think that they can construct a better ramp. Allow them to make some adjustments and collect more data. To ensure comparable data throughout the class, require each group to collect data from three ramps. Each mission team should collect data on ramps with height-to-length ratios of $1/3$, $1/4$, $1/5$, and $1/6$. Be certain that students release the cars from the same point on the ramps during the trials. Mark the release point with a piece of tape to ensure that it remains constant. Have each group predict which ramp will produce the highest speed.

The data should be organized as shown below, and the average speed should be calculated in feet per second.

Sample Data
Mission Team: Green

Ramp Ratio	*Trial*	*Distance (ft)*	*Time (s)*	*Avg. Rate of Speed (ft/s)*
$1/3$	1	8	4	2
$1/3$	2	8	5	1.6

Students are accustomed to miles per hour as a description of speed. Some may want to convert the results into miles per hour because that measure is the only everyday reference they have for speed. To give students a comparison for this experiment, help them convert a slow speed, say, 20 miles per hour, to feet per second.

To close this experiment, have students compare data about the ramps and the speed of the cars. Also have them compare their results with their earlier predictions.

Teaching Note

In reality, a car rolling down a ramp constantly accelerates; thus, the speed increases as the car nears the end of the ramp. As it continues across the table or floor, the behavior of the car accelerating down the ramp changes because of gravity and friction. Ideally, the car would travel at a constant velocity. As a simplification for middle school students, this activity uses the concept of average rate of speed over the distance traveled.

Teaching Tip

If the class has reached an appropriate point in the curriculum for use of the Pythagorean theorem, allow students to calculate the lengths of the sides of the triangle formed by the ramp.

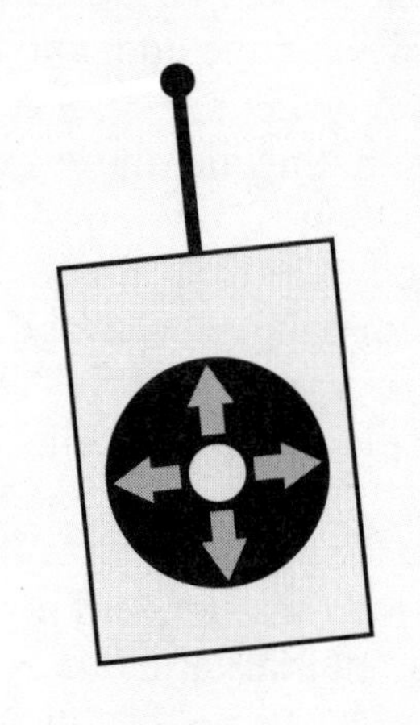

A remote-control car

Conducting the Experiment Using Battery-Powered Remote-Control Cars

At least one battery-powered remote-control car is needed for this experiment. These cars come in many varieties, including trucks, stock-car racers, dragsters, and so on. These cars can be very quick. On a straight track, the car can go quite a distance in 5 to 8 seconds. For this reason, the first test of speed should be done on an oval or circular track of known length. The closed track can be constructed indoors in a gymnasium, in a cafeteria, or on a stage or outside in the school yard.

Describe a track made of a rectangle with a semicircle on each end. In mission teams, have students design their own tracks. Students' experiences with formulas for perimeter and circumference may determine the steps they take in designing the track. Some students might develop a track shape, then measure the distance.

Alternatively, students can select a track length, then design a track of that distance. A circular track is acceptable and can be "drawn" by students using as the radius a length of string with a piece of chalk tied to the end.

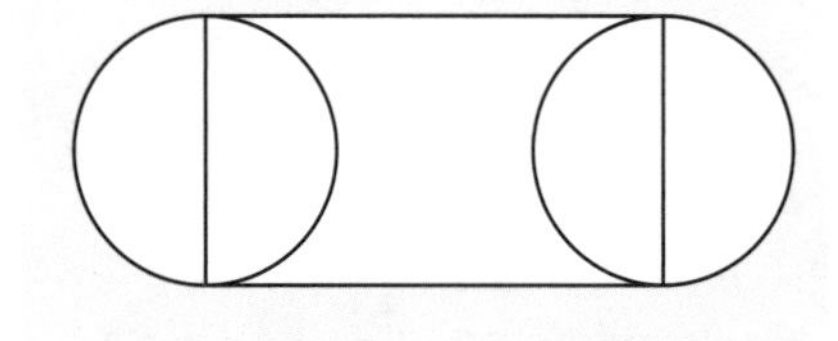

A pattern for a race track

The track designs from all the groups should be displayed and discussed. For the first trial, select an appropriate track for the available space. If more than one design is usable, set up more tracks to collect data. Define the track by placing traffic cones inside the track at such points as the beginning of turns and straightaways.

To conduct the experiment, each mission team should collect data on the time required to complete one lap around the track. This time will be used to compute the average speed of the car in one lap. Members of the mission teams can play the roles of starter, timer, recorder, and driver. Each student may want to "drive." Four trials for each team may be necessary to involve all students.

The data from the experiment should be collected in a table by each mission team, as shown below.

Sample Data
Mission Team: Blue

Track	*Driver*	*Distance (ft)*	*Time (s)*	*Avg. Rate of Speed (ft/s)*
1	Sue	54	6	9
1	Tina	54	18 (spin out)	3

The closing of this activity is interesting. Because the data will vary greatly from one trial to another, a class discussion will be necessary about how to report central-tendency data from the mission teams to the class. What approach is most advantageous if the teams are competing? What result can be used to be the most consistent across teams? What data best describe the speed of the battery-powered cars? Regardless of how the data are reported, students gain an intuitive and informal understanding of speed through these experiments.

Conducting the Experiment Using Toy Rockets

All rockets must be flown outdoors. Plan carefully to ensure that all safety precautions are followed. Use this opportunity to discuss the importance that NASA places on safety and to assure the class that the same considerations will be applied in this experiment. All forms of rockets are sold with specific directions related to construction and safety concerns. Read these directions carefully, and make students aware of their importance.

Toy rockets are available in toy stores in several forms, including compressed-air, water-powered, and solid-fuel rockets. Some students may own such rockets or may have had experience flying them in a hobby club. The discussions here are based on the use of solid-fuel rockets because they require the most preparation and involve significant safety concerns.

The speed of rockets is more difficult to measure than the speed of toy cars because rockets travel vertically. Accurately measuring the distance traveled is not possible with the tools at our disposal. Most students in middle school are not ready to measure the height of an object indirectly. For the majority of students, benchmark heights, such as that of the school building, a tree, or the flagpole, help them estimate the height of the rocket flight. Algebra students in middle school, however, may be able to understand the basic trigonometry necessary for indirect measurement.

In some classes, students can bring in rockets from home. In other classes, no one has rockets to share. When rockets and related equipment must be purchased, the cost of purchasing one rocket for each student is prohibitive. This experiment can then become a whole-class activity.

Student and parent volunteers can begin building the rockets and associated equipment about two weeks before the anticipated data-collection event. This phase needs to be done only for the first school year in which the experiment is conducted. In subsequent years, replacing the fuel is the sole preparation required.

For the data-collection event, group students into mission teams. All teams may collect data simultaneously. Two mission team members watch and estimate the height of the rocket flight, and two measure the time. Even with a warning about the shortness of the flights, students may miss the start or forget to push the stop button on the stopwatch when the rocket reaches its highest point. Usually, one of the two students gets both tasks done. Because the duration of the flights is short,

NCTM Teaching Standards

The teacher of mathematics should create a learning environment that fosters the development of each student's mathematical power by ... using the physical space and materials in ways that facilitates students' learning of mathematics.

(NCTM 1991, p. 57)

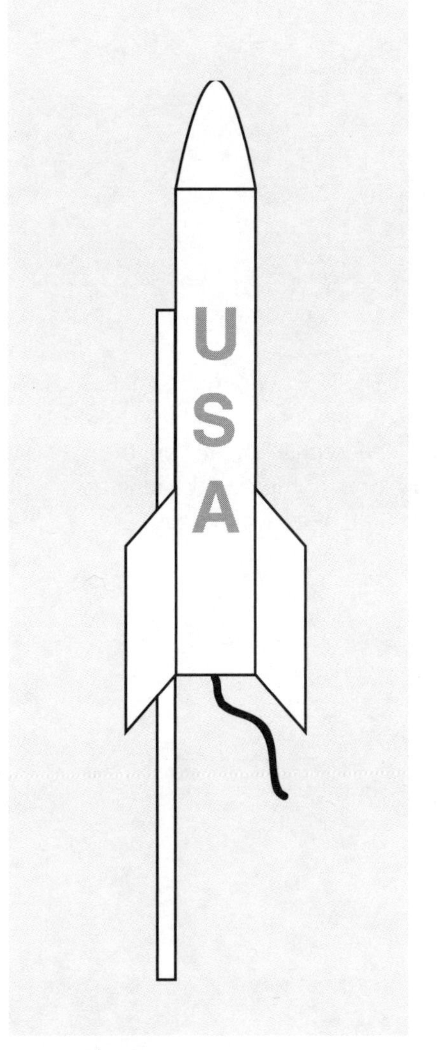

A toy rocket

variance occurs in the measures of time even if all students start and stop their watches appropriately. The jobs can rotate among team members from trial to trial. To get reasonable data, at least three rockets should be fired in each class. For both the height and the time, the two team members must agree on a value for the data chart.

Sample Data
Mission Team: Red

Rocket	Trial	*Distance (ft)*	*Time (s)*	*Avg. Rate of Speed (ft/s)*
1	1	60	1.5	40
2	1	75	2	37.5

Data usually vary among the mission teams. As a class, discuss how to summarize the data. Make a line plot of data for each rocket and trial. Discuss how to handle data that seem to be far from the others and how to use central tendency to describe those data that seem consistent. Try to come to a class agreement on an estimate for the average speed of the rockets in feet per second. Another rocket activity, "Fizzy-Tablet Rockets," from *Mission Mathematics II: Grades 3–5* (Hynes and Hicks 2005), may be a useful supplement to this activity. In that experiment, antacid tablets are used as fuel to launch film-canister rockets. The duration of the film-canister flights is short, but this activity can be used as an alternative to measure speed.

Concluding the Activity

The closing of this part of the activity varies somewhat, depending on which experiments are completed. However, each experiment indicates that summarizing the data should be part of the closing. Students should include these summaries in their logbooks, along with accounts of the mathematics they used and learned in the experiments. Of course, students should also record what they learned about speed from their experiments with toys. They should begin thinking about the speeds of running animals, swimming creatures, and various vehicles. Asking students to write two sentences about speed in everyday life will give insight into their current understanding of speed.

Concluding the Activity

In "An 8-Second Trip," students collected data about their speeds for walking 100 feet. They then collected data about the speeds of selected toys. These data can be entered on a class chart and discussed. The table included on data sheet 6 in the appendix (page A-8) contains data about the top speeds of some vehicles. If more data are desired about actual speed records, the *Guinness Book of World Records* (Young 2002) may be a useful reference. Students may be challenged to compare the two sets of data. Because one source shows data in feet per

NCTM Teaching Standards

The teacher of mathematics should orchestrate discourse by ... deciding when to provide information, when to clarify an issue, when to model, when to lead, and when to let students struggle with a difficulty.

(NCTM 1991, p. 35)

NCTM Teaching Standards

The teacher of mathematics, ... to enhance discourse, should encourage and accept the use of ... computers, calculators, and other technology.

(NCTM 1991, p.52)

second and the other presents information in miles and hours, comparing the data can be problematic.

The problem is how to convert the units in the data tables to allow comparisons. Given that most of the earlier data are in feet per second, encourage students to convert the data on sheet 1 to those units.

$$40 \text{ ft/s} \times 60 \text{ s/min} = 2{,}400 \text{ ft/min}$$
$$2{,}400 \text{ ft/min} \times 60 \text{ min/h} = 144{,}000 \text{ ft/h}$$
$$144{,}000 \text{ ft/h} \times 1 \text{ mi}/5{,}280 \text{ ft} = 27.27 \text{ mi/h}$$

This conversion offers an opportunity to encourage the use of technology to help solve a problem. Perhaps the most appropriate technology for this problem is a computer spreadsheet program. If the class can go to a computer laboratory where students can work independently or in pairs, the spreadsheet experience can be valuable in getting them to think algebraically.

The spreadsheet should have the following columns: Mode of Travel, Distance, Time, Distance in Feet, Time in Seconds, Speed in Feet per Second, and Distance in 8 Seconds. Students can enter the data in the first three columns from the data table. Through questioning, lead the class to propose how to convert miles to feet. Show students how to enter the multiplication of a constant, 5,280, and each number in the Distance column by using a formula. Additionally, students can copy and paste the formula throughout the Distance in Feet column.

Converting the time data to seconds involves a more complex formula. Including exponents in the formula may be appropriate if this topic has been covered in the curriculum. Whether the formula uses

Teaching Note

Another approach to the discussion is to use the maximum speed of animals. Crawling insects, mammals of the sea, birds, fish, reptiles, and land-based mammals can all be used to develop charts of distance, time, and rate of speed. Hands-on activities related to speed can be accomplished in classrooms using hamsters and turtles. Animals can be released near the middle of a circle and timed to determine how fast they reach the edge of the circle. Students might also hold a "Mutt Dog Derby" for their pet dogs. On a straight track, the dogs can be timed for a short race, then students can determine the average speed of each.

Top speed for various animals

exponents or the strategy of multiplying by 60 twice, both hours and minutes must be converted to seconds. Because this formula is more complex, it serves as a good example to show students how the spreadsheet's copy command is used.

This conversion can be accomplished without computers, using calculators or paper and pencil.

When they have completed the data table, challenge students to graph the results. Expect students to have some difficulty in making the graph. Because of the speed of the Space Shuttle, the range of the data is very large. The choice of a unit is difficult for representing a clumped group of data together with one extreme outlying piece of data. One remedy for this difficulty is to omit the Shuttle's speed and graph the remaining data. As a class, look at the graph and discuss what it would look like if the data were transformed into miles per hour.

NCTM Assessment Standards

Assessment should enhance mathematics learning.... [A]ssessments are learning opportunities as well as opportunities for students to demonstrate what they know and can do.

(NCTM 1995, p. 13)

ACTIVITY 3

Travel in the Solar System

This activity focuses on human travel in space. One problem associated with traveling in the solar system is the distance from one planet to another, but other problems also arise. A spacecraft needs fuel to make the long journeys in space, and humans traveling in space need food and water. The effects of microgravity, that is, near zero gravity, on humans over time are unknown. The probability of collisions with asteroids is uncertain, and many other aspects of long manned flights increase the complexity of space travel.

However, publicity about unmanned flights to the planets continues to raise the question of human space travel. Research is required to increase the probability that prolonged space travel for humans can be accomplished safely. One NASA project that will move us closer to space travel is the International Space Station, which will serve as a platform for many research agendas associated with living and working in space for long periods of time.

NCTM Standards

Instructional programs from prekindergarten through grade 12 should enable all students to—

Number and Operations

Understand meanings of operations and how they relate to one another

Compute fluently and make reasonable estimates

Measurement

Understand measurable attributes of objects and the units, systems, and processes of measurement

Apply appropriate techniques, tools, and formulas to determine measurements

Important Mathematical Ideas

Students apply measurement and computation skills to gain insight into numbers associated with distances in space.

Mission

Students plan a trip for humans to travel to a planet in the solar system.

Materials and Equipment

Data sheet 1 (see appendix, page A-8); "The Planets at a Glance" (see appendix, page A-6).

Launching the Activity

Begin the activity by engaging students in a discussion about humans traveling through the universe. In the movies and on television, students encounter science-fiction stories about traveling at the speed of light and beyond to cross entire galaxies in a matter of seconds. The experiences in "An 8-Second Trip" have reminded us that most of our travel speeds are quite slow. We have not come close to traveling at the speed of light. Although radio signals can reach Mars from Earth in a short time, the time required for a spacecraft from Earth to reach Mars is much longer.

From data sheet 1 on page A-8, use the data about the Space Shuttle to determine its speed in miles per hour.

Distance traveled by the Shuttle spacecraft
= 4,164,183 mi

Time to travel given distance
= 9 d 23 h 30 min = 9 d 23.5 h

9 d 23.5 h × 24 h/d = 216 h + 23.5 h
= 239.5 h

Given that $d/t = r$, we have

$$\frac{4,164,183}{239} = 17,386.98539 \text{ mi/h},$$

or approximately 17,400 mi/h

Teaching Note

For students who wish to think about the relative positions of two planets in their respective orbits, teachers may wish to refer to the activities related to scaling in *Mission Mathematics II: Grades 9–12* (House and Day 2005).

NCTM Teaching Standards

The teacher of mathematics should create a learning environment that fosters the development of each student's mathematical power by ... consistently expecting and encouraging students to ... work independently and collaboratively to make sense of mathematics.

(NCTM 1991, p. 57)

The teacher of mathematics should pose tasks ... [that] call for problem formulation, problem solving, and mathematical reasoning.

(NCTM 1991, p. 25)

Astronaut on the Moon with Earth in distance

Remind students that the Shuttle is not designed for travel among the planets. It is designed only for Earth-orbit tasks. However, its speed is helpful in judging the speeds of twenty-first-century spacecraft. After students have done the calculations, come to some agreement on an approximate speed for interplanetary travel. Assume that the agreement is about 50,000 miles per hour. This figure gives us a reasonable speed to use in thinking about space travel today. In the future, speeds will undoubtedly increase.

Using the data sheet "The Planets at a Glance," students can determine the distance from Earth to each of the other planets. This task is not trivial. The distances in the chart are given in millions of miles. To facilitate computation and estimation, students need to translate 67.2 million miles, the distance from the Sun to Venus, into its full numeric form, 67,200,000. Before thinking about traveling to Venus, students must remember that Earth is about 93,000,000 miles from the Sun. Use the mean distance from the Sun to specified planets to calculate each distance from Earth to the targeted planet.

To make this exploration manageable for middle school students, the activities are based on the assumption that the planets are aligned at their mean distances. Explain to students that in actuality, this greatly simplified situation is unlikely to occur.

Developing the Activity

The students' next task is to calculate the time required to travel to each planet on spacecraft that travel from 10,000 to 100,000 miles per hour. Group the students into their four-person mission teams. Ask them to complete a chart for travel to all the planets of the solar system at the speeds shown below. They should use the mean distance of each planet from Earth.

In a typical class, students groan at the prospect of completing a chart with eighty entries. The groans provide the opportunity to challenge the teams to think of strategies for reducing the amount of calculation required to complete the chart. When patterns are used to complete entries, the teams should record them. All students in the teams should fill in the chart.

NCTM Technology Principle

Technology ... influences the mathematics that is taught and enhances students' learning.

Time Required to Reach the Planets in Hours

mi/h	*Mercury*	Venus	Mars	Jupiter	Saturn	Uranus	Neptune	Pluto
20,000								
30,000								
40,000								
50,000								
60,000								
70,000								
80,000								
90,000								
100,000								

Discuss the patterns that teams used to complete the chart, and list them on the chalkboard. After all team members have shared how their patterns helped reduce the workload, ask the teams how many different patterns they used.

At this point, the class has a complete chart for travel to the solar system's planets, computed in hours of travel. To increase students' understanding of the meaning of these times, pose questions that require students to think about the practicality of space travel:

- If we travel at the approximate speed of the Space Shuttle, which planets can we reach in less than 10 years?
- How fast must we be able to travel to reach Jupiter in less than 10 years?
- Traveling to some of the planets at some of the indicated speeds would take more than a lifetime, which is

The Apollo 15 module

NCTM Assessment Standards

Although assessment is done for a variety of reasons, its main goal is to advance students' learning and inform teachers as they make instructional decisions.

(NCTM 1995, p. 13)

Teaching Tip

For further study, *Mission Mathematics II: Grades 9–12* (House and Day 2005) has activities that extend the ideas in this activity.

about 75 years. Which planets are too far away to be reached in a lifetime?

- We would like to make round trips. Traveling at 10,000 miles per hour, to which planets could we make round trips in one lifetime?

These questions should be posed to the mission teams. The teams should discuss the questions and agree on a team response. After students begin thinking about time questions, have each mission team make up questions for the class to solve.

This part of the activity may require one class period. The end of this discussion is a good place to break if necessary.

Developing the Activity

The terrestrial planets are the four innermost planets in the solar system: Mercury, Venus, Earth, and Mars. They are called *terrestrial* because they have rocky, compact surfaces like Earth's. Jupiter, Saturn, Uranus, and Neptune are known as *Jovian,* or "Jupiter-like," planets because they are gigantic when compared with Earth and have gaseous compositions like Jupiter's. Jovian planets are sometimes called the *gas giants.* Pluto is not a member of either group. Its composition is unknown, but it is probably composed mostly of rock, ice, and frozen gases.

Present the following scenario to students:

Because humankind wants to know more about each of our planetary neighbors, we need to plan our travel to the planets. Select one terrestrial planet and one Jovian planet. Plan trips to the two planets and to Pluto. Describe the speed of your spacecraft, as well as the time required to reach the planet, remain 1 Earth year to explore it, and return to Earth. You may assume that advances will be made in the development of spacecraft and that speeds of up to 50,000 miles per hour will be possible.

Launch day for all missions is 10 December 2004. On what date will you arrive at the targeted planet? On what date will you return from each mission?

An artist's rendering of *Pluto Express*

These questions require that students convert such time intervals as 10.2 years into years and days. When the conversion results in a part of a day, round the value to the nearest day. Students may not be familiar with thinking about a date in the year as having an ordinal value in relation to the year, as, for example, in the following scenario:

1 July is the 183rd day of the year. Locate a reference calendar in which the ordinal value is given along with the date. Remind students that they are not beginning with 1 January, because launch was on 10 December. Students will also need to keep track of leap years and use 366 days for each leap year spent traveling to or visiting other planets.

Extending the Activity

To this point in the activity, students have considered space travel from the perspective of what happens to the traveler on the journey, but while the space travelers are visiting distant planets, life continues on its usual course at home on planet Earth. Students should be familiar with this aspect of travel from their previous experiences. While they are away from home, life goes on; on their return, they need time to catch up on all the news and events. Occasionally, while they are gone an event occurs that has a profound effect on them when they return.

Group the students in their mission teams. Tell them to imagine that their team was sent on a mission to their selected terrestrial planet. They know their launch data and have computed the duration of their trip and the date of their return. Although NASA keeps crew members posted on the news, by the end of their mission the members of the traveling mission team will have missed many important events, both personal and public.

Have each mission team serve as the "ground crew" for a space-traveling counterpart. The ground crew's task is to debrief the astronauts on their return to Earth. Each ground crew makes a list of important events that the astronaut crew should know about on its return. Of course, the names and some of the events will be fictitious, but they should be plausible for the time that would have passed on the journey. Be certain to include the results of regularly occurring events, such as elections, not just glamorous or catastrophic events. Such personal events as graduations of family members should be mentioned, too. Outcomes of sports events, such as the Olympics, the Super Bowl, and the World Series, may be important to some students.

Each member of the mission team should write about one of the trips to a planet. The description should include the launch date, the

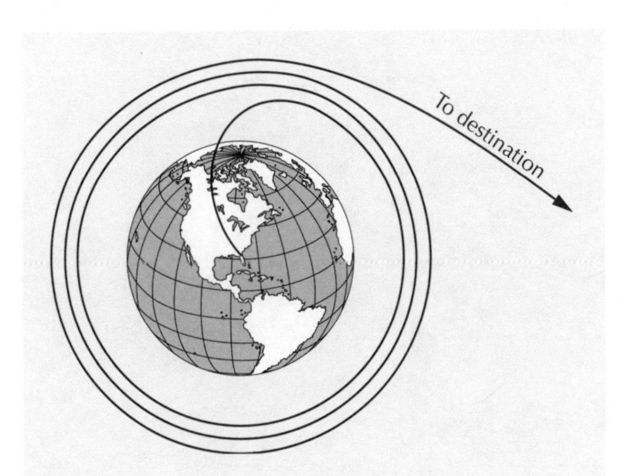

When a spacecraft is launched, it may orbit Earth a specified number of times before using the slingshot effect to transfer out of orbit.

Teaching Note

Students who are close followers of NASA space missions may realize that this activity simplifies some aspects of space travel and does not consider others. For example, a spacecraft is not aimed directly at a target destination in space in the way that a bullet from a gun is aimed, and the speed of a spacecraft is not uniform throughout its journey. Very sophisticated mathematics is required to calculate the intersection of two moving bodies, such as a spacecraft and a planet. This activity serves to introduce students to mathematical challenges for the future.

NASA Connection

www.stsci.edu/web/Copyright.html

heritage.stsci.edu/2001/24/

destination, the speed of travel, the elapsed time to reach the planet, the date of arrival, and the date the crew returned to Earth.

Each mission team should make a mission patch for one of its planned journeys. This patch can be attached to the logbook for this activity.

Students can present the briefings they have written for the returning astronauts. These reports could take many forms. Some mission teams may make time lines. Others may present their briefings in the form of a newscast. Some may use technology to support their presentations. Others may make scrapbooks. Do not place limits on their creativity.

All students should take time to reflect on the mathematics of this activity. The calendar, the time conversions, the distances in space, and the speeds required to complete space travel are all important concepts for students to think about as they construct their understanding of the world and the mathematics that describes it.

Teaching Note

Students who are ready to explore on the Internet may want to investigate the following sites:

www.jpl.nasa.gov/solar_system/planets.mars_index.html

www.jpl.nasa.gov/

www.nationalgeographic.com/ngm/0102/feature2/index.html

Space Rocks

Through the work of NASA scientists, who examine data related to the composition of the Moon and the planets, we continue to learn exciting things about our celestial neighbors. Helping students understand the roles of scientists and having them role-play those functions tends to improve their dispositions toward learning. This lesson illustrates techniques for studying other celestial bodies and motivates students to assume the roles of geologists.

Many students know that Apollo astronauts brought lunar rocks to Earth. They may have seen samples of these rocks exhibited at a NASA research center or a museum. More recently, rocks believed to be meteorites from the surface of Mars have focused increased attention on the scientific work of NASA geologists. Evidence found in the rocks raises the possibility that the environment of Mars may once have been hospitable to some forms of life.

On 4 July 1997, NASA's *Pathfinder* landed on Mars and released the six-wheeled robotic rover Sojourner. Sojourner is equipped with a scientific apparatus for conducting on-site analyses of Martian soil and rocks. This on-site capability allows scientists to study as many different rocks and types of soil as possible for as long as the rover and its instrument function. The *Pathfinder* mission found evidence to support the theory that water, which is necessary to life, once existed on Mars.

The *Pathfinder* mission will be followed by a series of robotic explorations of Mars to be launched through 2005 and beyond. The first of the explorers in the Mars Surveyor Program, the *Mars Global Surveyor,* entered the orbit of Mars in September 1997 with the primary mission of mapping the surface of the planet. The polar-orbiting spacecraft is designed to produce global maps of the surface topography and the distribution of minerals on Mars and to monitor Martian weather. Another orbiting spacecraft is *2001 Mars Odyssey,* designed to examine the composition of the surface of Mars, to look for water and ice, and to study radiation on the planet.

Using data systematically collected by orbiting spacecraft, two Mars rovers—Spirit and Opportunity—have been landed on the planet's surface to explore it from a closer vantage point. The mission of the rovers is to travel on the surface of Mars, recording and transmitting information for scientists on Earth. All this information is part of a long-term Mars exploration program that may involve peoples' landing on the Martian surface to collect data firsthand.

Preparing for human exploration of Mars requires the systemic collection and analysis of specialized data by NASA scientists. Such data are collected remotely and analyzed in laboratories back on Earth.

Collecting Mars Rock Data

The seven activities presented in this module are appropriate for all grade levels in middle school. All involve a high degree of student participation in hands-on activities. Thus, the amount of time needed to complete the module will vary; seven to ten class periods will probably be needed. If some of the mathematics required for the module must be taught, the time required may exceed ten class periods. These activities need not be presented on consecutive days. They can be interspersed among other mathematics activities over a longer time. The activities are as follows:

- "What Do Space Rocks Look Like?": Students identify attributes of "space rocks" collected by the class. Many of the attributes are revisited in subsequent activities as students take measurements and make comparisons.
- "How Can the Size of the Rocks Be Measured?": This activity addresses techniques for measuring attributes of objects having irregular shapes, in this instance, space rocks.
- "How Can the Rocks Be Measured with a Balance?": Students apply concepts from algebra to measure the masses of rocks using a balance with standard masses and other rocks and to make comparisons.
- "How Dense Are the Rocks?": The topic of density is explored using various balls and the space rocks.
- "How Hard Are the Rocks?": This activity focuses on one of the attributes of minerals: hardness. Students learn about this attribute and develop a scale of hardness, then apply the scale to their sets of space rocks.
- "Where Are the Rocks?": Students play the role of scientists in the field who are creating a site map of a research area. They use another mission team's data to re-create a research site by locating all rocks on the site.
- "How Big Are the Rocks on Mars?": Students describe and catalog space rocks using the notion of attributes. They learn to use indirect measurement and approximations to measure some attributes of the rocks.

All these activities require a collection of rocks. To get a good variety, students may need to bring in interesting rocks they have already collected or have found in their neighborhoods. If all rocks are collected from the same neighborhood, however, they may be quite similar, so students may have difficulty finding differences in their characteristics. If possible, order at least one set of rock specimens from a science supply house, or ask a science teacher to loan the school's geology kit.

For all the activities, group students into four-member mission teams. The team members simulate the work of scientists who examine rocks from the Moon or one of the planets. Students examine their space rocks and record the properties of the rocks on data sheets.

As they do for other modules, students use their creative ideas to link the mathematics they are learning with aeronautical themes by designing mission patches and logbook covers. Either or both of these activities can be continued throughout the modules. When students work in groups, some may complete their tasks before others and may need additional worthwhile learning activities. Assigning students to design mission patches and logbook covers is one way to use this extra time productively and extend opportunities for learning.

Pictures of a small pan balance or of a scientist collecting specimens in the field are common elements often found in patches for this unit. The inclusion of these items on their patches indicates the elements of the module that are important to students. A great deal of reflection about their learning occurs when students are deciding what to include in their patches.

Internet Connection

NASA mission patches can be found on the Web at the following site:

www.hq.nasa.gov/office/pao/History/mission_patches.html

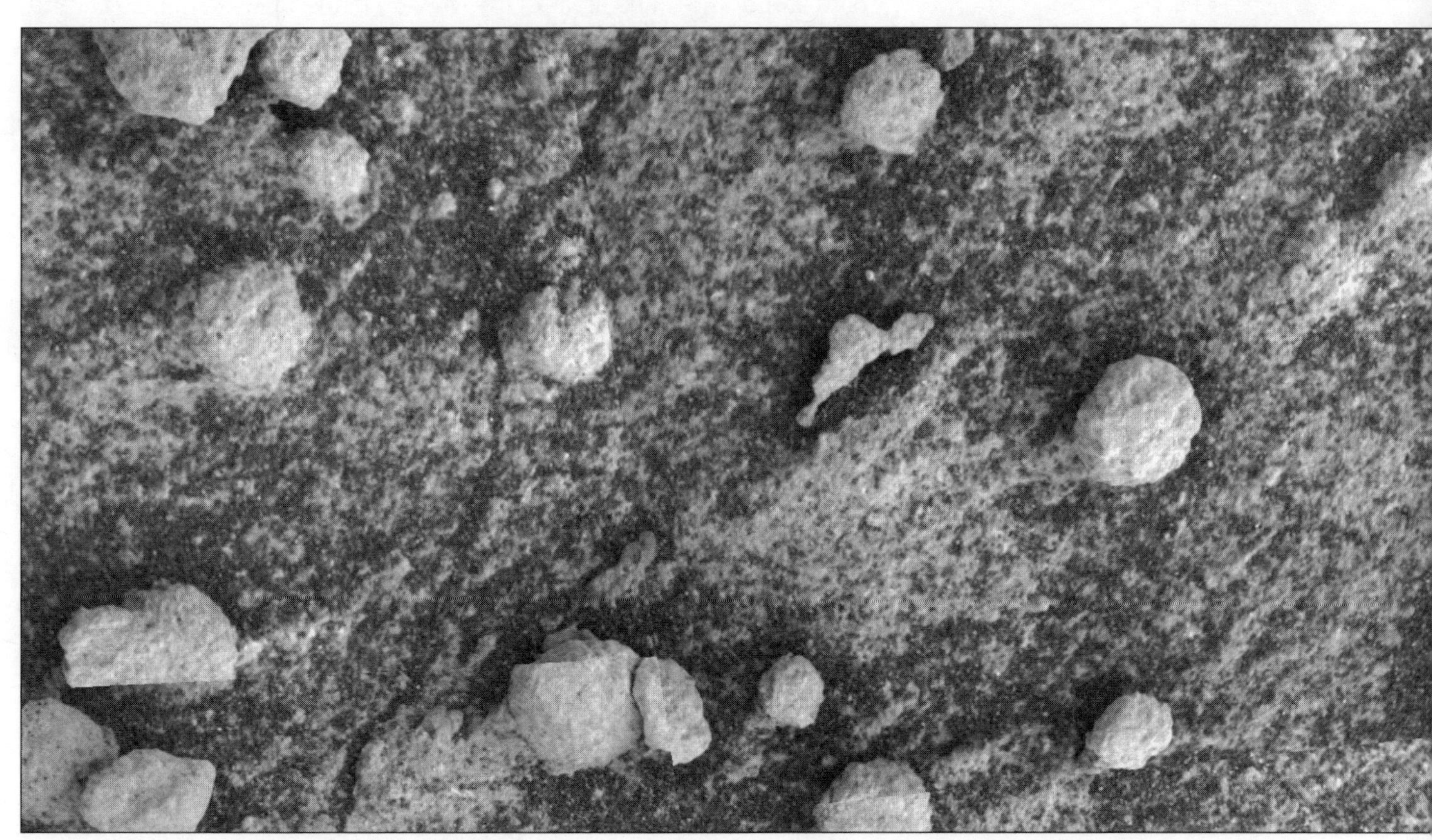

This image features the rock target dubbed "Bylot," acquired by NASA's Mars Exploration Rover *Opportunity* on August 9, 2004. The spherules shown here are less round than the "blueberries" seen in "Endurance Crater," perhaps because the minerals coating them are more resistant to erosion. Dark sand is partially covering the rock.

ACTIVITY 1

What Do Space Rocks Look Like?

NCTM Standards

Instructional programs from prekindergarten through grade 12 should enable all students to—

Measurement

Understand measurable attributes of objects and the units, systems, and processes of measurement

Geologists are scientists who study rocks and rock formations. They communicate about rocks by describing their attributes. Also, NASA Scientists expend a great deal of effort to determine the attributes of rocks. The rocks brought from the Moon and meteorites, which are rocks from space that have hit Earth, are of special interest to scientists.

Astronauts on the Apollo missions to the Moon returned with rocks that are still considered treasures of the world. They are kept in a controlled environment, and only certified professionals have access to them. Scientists have analyzed these rocks to catalog their characteristics, determine their mineral content, and examine them for fossil remains.

More recently, scientists found some suggestions of the possibility of primitive life while examining the remains of meteorites believed to have come from Mars. This discovery has increased the importance of future unmanned probes of Mars. *Pathfinder* is one of the probes designed to roam around specified areas of the Martian landscape, examine soil samples and rocks, and analyze the composition of this material.

In preparation for the activity, ask students to bring to class interesting rocks they have found in their yards or neighborhoods or on vacation trips. These can become part of the "space rock" collection.

Important Mathematical Ideas

Students determine quantitative and qualitative attributes of blocks and rocks. The attributes that are measurable are used to develop students' understanding of techniques for measuring attributes of irregular shapes.

Mission

Students use measurement and descriptive terms to describe the attributes of space rocks.

Materials and Equipment

Collections of rocks and sets of attribute blocks

Launching the Activity

Begin with a brainstorming session by the class. As in all such sessions, the teacher should accept all responses to the initial question as being genuine conjectures. After the brainstorming session, the list of conjectures can be prioritized and the more plausible ones can be retained for further study.

The question for this activity is "How have scientists collected rock specimens from outer space to study in laboratories on Earth?" Students might respond in the following ways:

- Some students may know about the Apollo missions.
- A few students may remember reading about the Mars meteorites.
- Some will say that aliens brought the rocks to Earth.
- Some will say that Space Shuttle astronauts brought rocks back as souvenirs.
- Some will say that we have no rocks from other celestial bodies.

Many additional creative responses to this question will be offered. Accept and record all responses, then have students rank the items on the list from the most likely to the least likely. This discussion presents an opportunity to talk about the space program and the exploration of our solar system. Students should learn the role of the Shuttle to remove any misconceptions that Shuttle astronauts visit the Moon or any of the planets.

Developing the Activity

Show students the rock collection, and compliment them for bringing in a wonderful collection of space rocks. Tell them that they will begin to study these rocks soon. Before studying these unfamiliar artifacts from space, however, the class will look at some familiar objects.

Show the class an attribute block, and talk about why it has that name. What are the attributes of the block? Each mission team should create a list of attributes and share the list with the class. The final class list will comprise typical attributes, such as color, size, shape, thickness, texture, weight, and so on. Some of these characteristics are measurable; some are merely descriptive. For example, weight is measurable but color is a descriptive term. In a subsequent activity, students develop a measurement scheme for the hardness of rocks. The suggestion of hardness as an attribute early in the module is a bonus for the teacher; however, if hardness is not included in the list, it can be introduced as a new attribute for the targeted lesson. Through class discussion, divide the overall list into separate lists of descriptive and measurable attributes.

Invite students to look at the rock collection they have created for their simulation. They should examine the space rocks and select one. Referring to the class list of attributes for the attribute blocks, ask students to select attributes that apply to their space rocks. Are other characteristics important for describing their rocks?

In their mission teams, students make a list of attributes that apply to their rocks.

Concluding the Activity

Develop a data sheet for recording the information about space rocks to close this part of the lesson. Mission teams should report their ideas

NCTM Teaching Standards

The teacher of mathematics should orchestrate discourse by—

posing questions that elicit, engage, and challenge each student's thinking;

listening carefully to students' ideas;

deciding what to pursue in depth from among the ideas that students bring up during a discussion.

(NCTM 1991, p. 35)

The teacher of mathematics should promote classroom discourse in which students ... try to convince themselves and one another of the validity of ... conjectures and answers.

(NCTM 1991, p. 45)

NCTM Assessment Standards

Students should be able to see the connection between instruction and assessment.

(Adapted from NCTM 1995, p. 22)

NCTM Teaching Standards

The teacher of mathematics should pose tasks that … represent mathematics as an ongoing activity.

(NCTM 1991, p. 25)

on attributes that might be important to record about the rocks. New attributes may be added to the previously developed list using attribute blocks. For example, some students may want to note whether a rock has fossils, or is shiny or dull, all of which are indeed important characteristics.

Each student should rewrite the list of attributes as a data sheet for recording information about the rocks in subsequent activities.

Extending the Activity

The class can create a database with a field for each identified attribute, then collect and enter data in subsequent activities.

ACTIVITY 2

How Can the Size of the Rocks Be Measured?

Important Mathematical Ideas

Students measure attributes of irregularly shaped objects using direct and indirect measurement techniques.

Mission

Students measure the surface area and volume of irregularly shaped objects to simulate the job of NASA scientists in examining rock specimens from other celestial bodies.

Materials and Equipment

A collection of rocks; sets of measuring instruments, including rulers, grid paper, string, graduated cylinders, and marked beakers; water; graph paper; small boxes; and log sheets (see appendix, page A-2)

Launching the Activity

Give each student one space rock from the class collection. Organize the class into mission teams. The teams' task is to record both descriptive and measurable data about the space rocks in their collections. Every student should record data about all the rocks in his or her team's collection. The data from this activity can be incorporated into the class database, which allows students to sort the data according to attributes. Additionally, students can calculate mean measures for the collection.

Developing the Activity

In the process of conducting this activity, the teacher may need to gather all members of the class to talk about how to measure certain attributes. Two attributes, surface area and volume, may be somewhat difficult for students to measure, and they may need guidance in making estimates for these attributes. Students may know formulas or even some techniques for measuring objects that have regular shapes, such as attribute blocks, but these procedures are ineffective for measuring objects with irregular shapes. If some mission teams find good ways to estimate these measures, have them share their ideas with the class.

Use an attribute block to review the definition of surface area and the units of measure associated with this attribute. Try to help students think about surface area as covering the object using graph paper or aluminum foil with no overlaps. This technique can be used to approximate the surface area of irregularly shaped objects.

The discussion of surface area can include a series of approximations. Begin with a box in which the rock just fits as the first approxima-

NCTM Standards

Instructional programs from prekindergarten through grade 12 should enable all students to—

Measurement

Understand measurable attributes of objects and the units, systems, and processes of measurement

Apply appropriate techniques, tools, and formulas to determine measurements

tion. Hold the box containing the rock up to the light to allow students to see the gaps between the rock and the box. Eliminate one gap by carefully cutting off or "denting in" one of the corners. Continue this process to show a series of closer approximations. This exercise can serve as an introduction to the concept of limit.

The formulas for volume cannot be applied to this task. As one student said, "You can't do this; there isn't an *l*, *w*, or *h*." A new technique is needed. Use this opportunity to highlight one of the advantages of the metric system.

In the metric system, volume is measured in such units as cubic centimeters, or cm^3. Coincidentally, the amount of water that fills a 1-cm^3 container is 1 mL. Place a rock into a graduated cylinder or beaker with a known amount of water; the amount of water in mL displaced by the rock is its volume in cm^3.

For example, use a 100-mL graduated cylinder filled with 50 mL of water. Put a small rock in the graduated cylinder. If the water rises to the 60-mL mark, the rock displaces 10 mL of water, which has a volume of 10 cm^3. The volume of the rock is 10 cm^3.

Concluding the Activity

Students from each mission team report to the class about the attributes of their most interesting rocks. Different students can talk about the different attributes and ways they measured the rock for some of the attributes. From this activity, the preliminary class list of attributes may be augmented. Some students may have investigated an attribute not included on the initial class list. The mission teams may need to meet again to record more data about their rocks.

All students should write in their logbooks about the mathematics used in this activity.

ACTIVITY 3

How Can the Rocks Be Measured with a Balance?

Important Mathematical Ideas

Students develop some notions of algebra by considering the mass of the space rocks. Relationships between masses of rocks are described using algebraic equations.

Mission

Students measure the mass of objects using a balance.

Materials and Equipment

A collection of rocks, balance scales, standard masses, and log sheets (see appendix, page A-2)

Launching the Activity

Mass can be thought of as closely related to weight but with some differences. Make sure students understand that mass is a property of matter, which is defined as any physical substance. In contrast, weight is a measure of the force that gravity puts on matter. Because of a perceived absence of gravity, objects in orbit appear to be weightless, but these objects nevertheless have the same mass they would have on Earth.

Give each mission team a pan balance and some standard masses to measure the mass of the space rocks. The standard masses used can be commercially made centicubes that have a mass of 1 gram (1 g) or any other standard masses. Central to this activity is the preparation of the materials. Before the activity, each team should carefully examine its rock collection. Teams should be given too few standard masses to determine the masses of its heavier space rocks. The point of this activity is to compel students to use smaller rocks and standard masses to find the masses of larger rocks.

Developing the Activity

Tell the teams that their task is to describe the mass of each of the space rocks in their collections. They may use only the materials they have in the activity laboratory package. Students should discover that they do not have enough standard masses to balance the heaviest rock. The typical first action of the students is to ask for more standard masses, which should not be given. At this point, encourage students to think of ways to balance the rocks with just the materials found in their laboratory package.

NCTM Standards

Instructional programs from prekindergarten through grade 12 should enable all students to—

Algebra

Represent and analyze mathematical situations and structures using algebraic symbols

Use mathematical models to represent and understand quantitative relationships

Some students in each team should realize that a smaller rock can be used with standard masses from the provided set to determine the masses of the larger rocks. Students can combine a smaller rock with standard masses to balance a larger rock.

Have students write a sentence that describes the mass of a large rock by using a smaller rock and standard masses. For example, one team might write, "The mass of the large reddish rock equals the mass of the small brown rock plus three units of mass." This statement can be translated into an algebraic sentence, as follows:

y = the mass of the large reddish rock,

x = the mass of the small brown rock, and

$y = x + 3.$

Other rocks will have different masses. For example, another rock may have a mass of $x + 7$. Students can compare these two rocks to see that $x + 7$ is greater than $x + 3$.

Students discover that a smaller rock can be used with standard masses to determine the mass of a space rock.

Concluding the Activity

Challenge teams to express the mass of each of their rocks using standard masses or other rocks together with standard masses.

Eventually, students should substitute the measures for the masses of the smaller rocks to solve these equations. The masses of the rocks will be used to explore density in the activity "How Dense Are the Rocks?"

In learning to express the mass of rocks algebraically, some groups may complete their work before others do. Writing algebraic sentences is a challenge for some students and second nature to others. Developing the mission patch and designing the logbook cover for the module can serve as additional learning experiences for students who complete assigned work early.

Extending the Activity

Have teams exchange rock collections and make new algebraic sentences to express the masses of the space rocks. Once students become accustomed to writing algebraic sentences in this activity, ask them to write in their logbooks about this experience in using algebra.

ACTIVITY 4

How Dense Are the Rocks?

Important Mathematical Ideas

Density is a difficult concept for students to grasp. The relationship between mass and density is explored in this activity.

Mission

Students determine the density of regular and irregular objects.

Materials and Equipment

A collection of rocks, various balls, rulers, weighted thread, balance scales, and standard masses

Launching the Activity

Density is a measure of the amount of matter packed into a given space. Mathematically speaking, it is the average mass per unit volume. The more closely packed the molecules, the higher the density of the material. In general, planetary bodies in our solar system are composed of iron, rock, ice, liquids, gases, and on Earth, organic materials. This activity focuses on techniques to determine the densities of irregularly shaped objects, the space rocks.

Density can be a difficult concept for students to grasp. Using an exaggerated example may be helpful in demonstrating the difference between mass and density.

Give each team two balls of varying sizes. Volleyballs, baseballs, tennis balls, and Ping-Pong balls will all serve the purposes of this activity. Make sure that some groups have larger, lighter balls, such as Styrofoam balls from craft stores.

The equation for density is mass (g)/volume (cm^3). Students may need to be reminded that the formula for finding the volume of a sphere is $(4/3)\pi r^3$. Tell teams to find the volumes for each of their spheres. Students may also use the masses and balances from the previous activity to find the mass of each sphere. Students may need to use the masses of the space rocks, determined in the previous activity, if they do not have enough standard masses to measure their spheres.

At least one team should have a smaller sphere with a higher density than a larger sphere in the same collection. Ask for a volunteer to explain how one object can be both smaller and denser than a larger object. Ask a team to state the density of one of their spheres, along with the method they used to find it. Each of the densities for the spheres should be determined before this activity to enable comparisons of the teams' results with the actual densities.

NCTM Standards

Instructional programs from prekindergarten through grade 12 should enable all students to—

Number and Operations

Understand meanings of operations and how they relate to one another

Algebra

Represent and analyze mathematical situations and structures using algebraic symbols

Use mathematical models to represent and understand quantitative relationships

Measurement

Understand measurable attributes of objects and the units, systems, and processes of measurement

Apply appropriate techniques, tools, and formulas to determine measurements

Developing the Activity

This point in the activity presents a good opportunity to talk about human error and accuracy in measurement.

Ask students to suggest other pairs of objects in which one object is smaller but denser than the other. Students might suggest a cotton ball and a marble, a baseball and a beach ball, or a book and a box of cereal.

Ask the mission teams to use the volume and mass data they collected in the previous activity to calculate the densities of the rocks in their collections.

Have teams create graphs to display the densities of their rocks. One or more of the descriptive attributes can be used to identify the rocks in the graphs. Students might choose to name the rocks by color, texture, size, or shape or a combination of two or more attributes.

These graphs can be displayed around the classroom or included in the students' logbooks.

Concluding the Activity

The concept of mass is difficult for middle school students to comprehend. Teachers should not be too concerned if students seem to understand the concept one day but are confused about it the next. In Piagetian terms, many middle school students have not reached the level of maturation necessary to "conserve" the concept of density.

Extending the Activity

To introduce a technology connection with this lesson, set up an LCD panel or a computer and projector to allow all students in the class to participate in building a database. Students can identify the fields for the data and decide which statistics are important for summarizing the data.

ACTIVITY 5

How Hard Are the Rocks?

NCTM Standards

Instructional programs from prekindergarten through grade 12 should enable all students to—

Algebra

Use mathematical models to represent and understand quantitative relationships

Data Analysis and Probability

Formulate questions that can be addressed with data and collect, organize, and display relevant data to answer them

Important Mathematical Ideas

Students develop and use a descriptive and quantifiable scale of hardness for rocks. For most students, this experience is one of their first in developing a valid scale for measuring some attribute of an object.

Mission

Students develop a scale of hardness, similar to the one that scientists use for minerals, to further describe the space rock collection.

Materials and Equipment

A collection of substances with varying degrees of hardness, such as a piece of chalk, a penny, a dime, a wooden object, and so on; boxes; a collection of rocks; and log sheets (see appendix, page A-2)

Launching the Activity

One task of scientists is to develop ways to describe what they encounter in the laboratory. Geologists in the space program, who examine rocks from other celestial bodies, must describe specimens in terms that fit in the reference system of Earth geology. One of the attributes of minerals that is important to scientists is hardness. Hardness is described as the resistance that a smooth mineral surface offers to scratching. Some minerals, such as talc, are very soft; in contrast, a diamond is very hard. In fact, a diamond can be scratched only by another diamond. Although an established scale of hardness exists, this lesson requires students to use common objects from their environment to establish their own scales, then test the hardness of rocks in the class collection.

A scientist determines the hardness of a given rock by trying to scratch one of its smooth surfaces with another material of known hardness. In 1812, Friedrich Mohs, a geologist, proposed that relatively common materials could be used to create a scale of hardness. The scale is presented in order of increasing hardness: (1) talc, (2) gypsum, (3) calcite, (4) fluorite, (5) apatite, (6) orthoclase, (7) quartz, (8) topaz, (9) corundum, and (10) diamond. Each mineral in the scale can scratch those with lower numbers but cannot scratch those with higher numbers.

Begin the discussion of hardness in class by defining the term. *Hardness* means the resistance to scratching by another object.

An example of the test of hardness that is used nearly every day in the classroom is "scratching" the chalkboard with chalk. As a result of trying to scratch the chalkboard, writing appears on the harder surface.

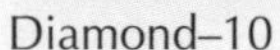
Diamond–10

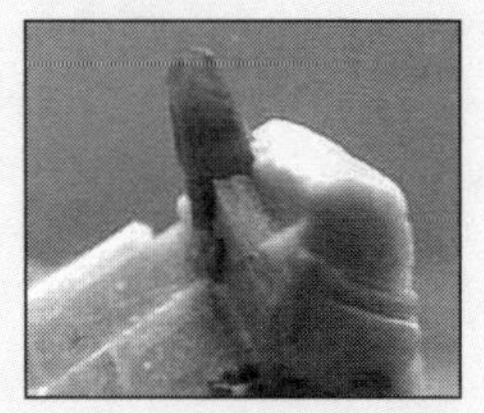
Corundum–9

Topaz–8

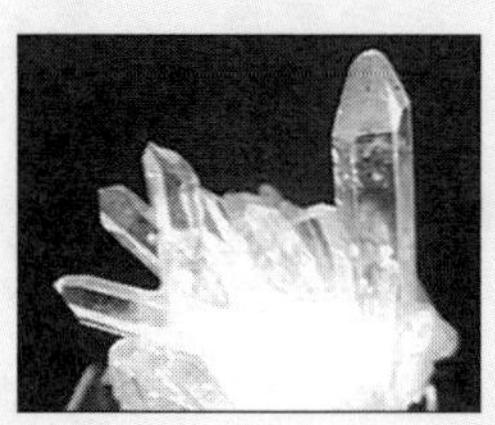
Quartz–7

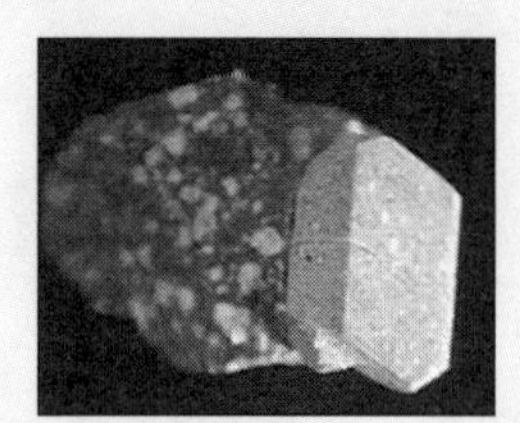
Orthoclase–6

Some minerals are relatively soft. The diamond is the hardest mineral on Earth.

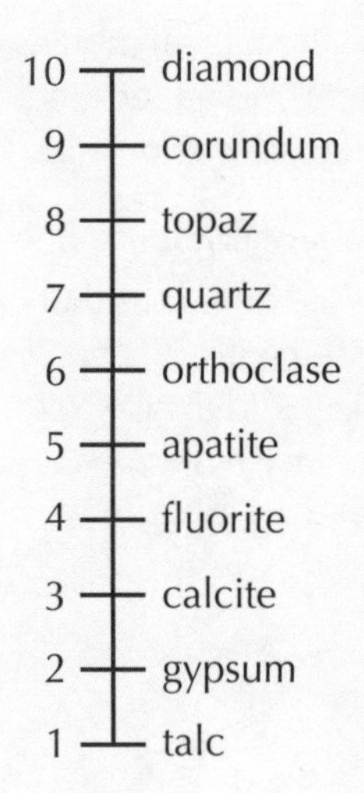

Moh's scale of hardness

The softer material—chalk—rubs off on the harder one—the chalkboard—and the softer material becomes smaller in size.

Ask the class to think of other instances in which the surfaces of two objects are rubbed together and one rubs off on the other. Make a list of the ideas generated by the class. Some of these ideas may lead students to include certain materials in their scales. For example, chalk is a good material for the students' hardness scales.

In giving examples of two surfaces rubbing together, students are likely to include drawing with chalk on the sidewalk, and some may say that chalk is softer than cement. Some students may have rubbed a penny on different surfaces and may wish to share those experiences.

Through discussion, rank the objects that students mention in the descriptions of their experiences with the hardness of objects. Rank these objects as scientists have done, from softest to hardest. The result might be a list similar to the following:

(1) Chalk

(2) Fingernail

(3) Penny

Confirm the ranking by demonstrating the scratch test. Does a fingernail scratch the penny? Will a fingernail scratch chalk? Each test of hardness should be demonstrated in front of the class.

Once the ranking is confirmed, assign numbers to each object on the list. For example, in the foregoing list, chalk has a hardness of 1 and a penny has a hardness of 3.

Developing the Activity

Next have students form their mission teams to examine the hardness of their space rocks. They should use the same rocks they previously described in "How Dense Are the Rocks?" The data collected on hardness should be added to the students' data sheets for their rocks.

Place as many boxes at the front of the room as the number of categories in the class hardness scale. The boxes should be labeled to indicate the levels of hardness. Ask a student from each mission team to place the rocks studied into the boxes according to their hardness. Dis-

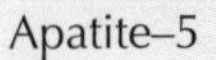
Apatite–5

Fluorite–4

Calcite–3

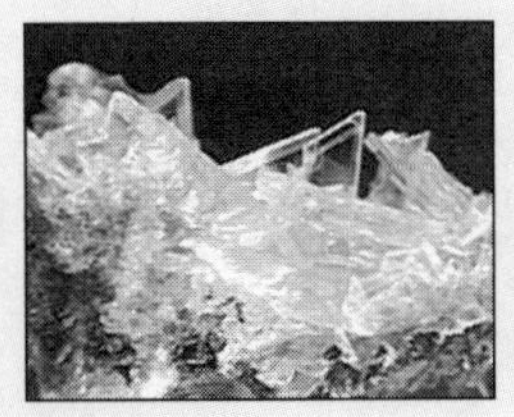
Gypsum–2

Talc–1

Talc can be found at www.minersoc.org/pages/gallery/minpix/talc/talc.html. All other minerals are copyrighted: © 1995, 1996 by Amethyst Galleries, Inc., mineral.galleries.com.

cuss the distribution of the rocks. Count the number of rocks in each box, and record the totals on the chalkboard.

Have students create bar graphs in their logbooks to show the distribution. Additionally, ask students to write about their experiences in formulating a scale for measuring this attribute.

Concluding the Activity

Hardness is a favorite theme for mission patches. Students may include messages about graffiti on their patches. This connection among mathematics, science, and social studies is meaningful to students and may also be reflected in their logbooks.

Extending the Activity

The distribution of rocks on the hardness scale may not include many soft or hard samples. Challenge students to locate rocks or other objects that can be tested for hardness and placed in the boxes for the softer and harder levels of the scale. This activity may be continued for several days or weeks.

Some students may want to know whether scientists actually use this procedure to measure hardness. Have them do research in the library, in the media center, or on the Internet to learn more about the scale of hardness. They should record the results of their research on the "Space Rocks" log sheet.

Curriculum Connection

The discussion about chalk may present an opportunity to make a connection with the social studies curriculum and the topic of responsibility. Talk about students' responsibility to keep their neighborhoods and schools free of graffiti. Scratching a name or a saying on someone else's property is disrespectful. People may harm trees by carving in the bark. Students should be made aware that these types of activities are harmful to other people and the environment.

ACTIVITY 6

Where Are the Rocks?

NCTM Standards

Instructional program from prekindergarten through grade 12 should enable all students to—

Number and Operations

Understand numbers, ways of representing numbers, relationships among numbers, and number systems

Geometry

Specify locations ... using coordinate geometry

During the collection of data, field scientists, such as biologists, geologists, archeologists, and others, try to keep track of where objects are found. The location may be an important clue to the nature of a plant species or the origin of a rock. For this reason, scientifically interesting areas are mapped carefully and locations of interesting objects are cataloged. Scientists sometimes re-create a scene in the laboratory to allow it to be studied in more detail. Space geologists use these techniques, too.

Data maps are essential to NASA scientists when they retrieve data from planet explorers, such as *Spirit* and *Opportunity*. Knowing the location of a rover designed to collect data about rocks and soil on Mars is important. Scientists may want to send another rover or a robotic expedition to the same site to confirm the existence of a significant geological find. Unless they know the exact location of the site and the objects in the field, verifying their finds is difficult.

As we all know, appearances can be deceiving, and the appearance of the surface of Mars is no exception. A view of the surface from a distance reveals little about the composition of the underlying rock because the planet is covered with dust, the product of billions of years of erosion by the Martian wind. In some places, rock is exposed, but in others, wind-blown dust has masked underlying features. Near the north pole, for example, the dust has been blown into dunes.

This meteorite, a basalt lava rock nearly indistinguishable from many Earth rocks, provided the first strong proof that meteorites could come from Mars.

Our knowledge of the composition of the underlying Martian surface took a leap forward in 1997 when NASA's *Pathfinder* spacecraft landed in the Xanthe Terra region of the northern hemisphere. On board was Sojourner, *Pathfinder's* small rover and the first instrument capable of analyzing soil and rocks in situ to be used on Mars.

"Before the *Pathfinder* mission, a general consensus held that the Martian surface is made from volcanic lava [mafic], like the Martian meteorites. But the first measurements by *Pathfinder* showed that the rocks contain the whitish mineral feldspar [felsic], rather like the Earth's continental crust," says Heinrich Wänke of the Max Planck Insitut für Chemie, Mainz, Germany. This organization designed and built the APX-spectrometer (alpha-proton X-ray spectrometer) on board Sojourner. Mafic rocks are high in magnesium and iron and are thought to derive from pristine mantle material; felsic rocks are rich in silicates, potassium, and sulphur but

A moon rock

low in magnesium and are thought to derive from rock that has undergone subsequent processing since the planet's formation.

Taking into account evidence from the Martian meteorites, *Pathfinder* data, and observations from orbiting spacecraft, planetary scientists have deduced that the composition of the low-lying plains in the northern hemisphere is felsic and that of the ancient highlands in the southern hemisphere is mainly mafic. This division corresponds well with the theory that the northern lowlands have a relatively recent volcanic origin, whereas the south exhibits a primordial crust.

Space missions so far have given us the general picture. The next step is to map the surface composition of Mars with greater accuracy. "We want to know the iron content of the surface, the oxidation level of the iron, the hydration of the rocks and clay minerals, the types of silicates present, and the abundance of nonsilicate materials, such as carbonates and nitrates," says Jean-Pierre Bibring of the Institut d'Astrophysique Spatiale, Orsay, France. Bibring is the principal investigator for the Infrared Mapping Spectrometer (OMEGA), an instrument on *Mars Express.* OMEGA, with the help of two other *Mars Express* instruments, the HRSC (high resolution stereo camera) and MARSIS (Marine Satellite Information Service), will map the composition of the Martian surface to this level of detail.

Important Mathematical Ideas

Simulating actual tasks of scientists motivates students to participate in their mathematics classes. Using the mathematics that scientists use in the laboratory and the field makes learning meaningful. In this lesson, students make a map of a field site by using the mathematics of coordinate geometry.

Link with Other *Mission Mathematics* Topics

Geologists and other field scientists are beginning to use Global Positioning System (GPS) units to record the locations of discoveries made in their work. Information on GPS can be found in the "Tracking Migrating Whales" module in this book and in "Finding Our Way" in *Mission Mathematics II: Grades 9–12,* (House and Day 2005).

Mission

Students apply coordinate geometry to make and use a data map for a simulated geology expedition.

Materials and Equipment

Yardsticks or metersticks, rulers, graph paper, rocks or colored counters, prepared sand trays or outdoor "field sites," and log sheets (see appendix, page A-2)

Launching the Activity

This activity can be conducted indoors or outdoors. Advance preparation is required for either location.

Indoors, sand-tray field sites must be prepared for each mission team before class. The sand trays can be made from shallow cardboard boxes that have been lined with plastic to keep the sand in the box. In each box, six to eight counters in different colors are placed in the sand to simulate rocks. The locations of these "rocks" should be generally random, but a few should be placed in each quadrant of the box.

Because all four quadrants of the Cartesian plane are used to develop the map, an understanding of integers is important for this task. If students have not studied integers and the Cartesian plane, limit the activity to the first quadrant.

Outdoors, field sites must be marked off before class. Yardsticks or metersticks can be used. Sites 1 yard or 1 meter square are sufficient for the activity. As in the indoor activity, six to eight "rocks" are placed in each field site and distributed in all four quadrants. For the outdoor activity, rocks from the class collection may be used.

Assemble the mission teams, and give them instructions before assigning their field sites.

Before the activity, tell students not to touch or move any of the rocks at the field site until given clearance. Invite them to look at the site carefully and discuss with their team members the descriptions of the rocks and their locations.

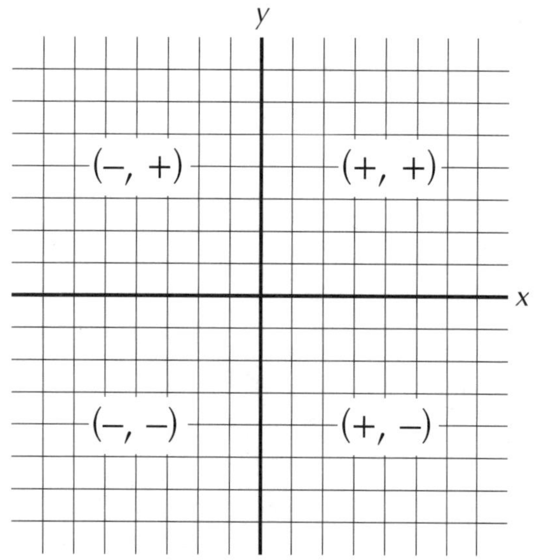

y

(–, +) (+, +)

x

(–, –) (+, –)

The four quadrants of a coordinate grid

Developing the Activity

Give each team a piece of graph paper. Ask teams to create a scale for a "map" of the site to be shown on a coordinate grid. Because the sand trays or outdoor grids are larger than a piece of graph paper, students must use proportional reasoning to create their maps. Note that for most middle school students, using a grid showing all four quadrants of the coordinate plane is appropriate. Some students may want to draw the axes in the sand. Doing so is permissible, but after creating their maps, students should rub out the axes for the next task.

Discuss the importance of accuracy in making the maps. Scientists who make these same types of maps understand that other scientists may want to re-create the site for further study. If the data are not accurate, the sites cannot be accurately re-created.

The data for this simulation should include a descriptive name for each rock or colored chip, together with its location. This information should be rewritten as a data table. Students should also include a scaled map of the location of the rocks as part of the data.

A re-creation of the sites will be simulated in class. After each mission team creates a map of its site and a data table of ordered pairs, the rocks are removed from the site, and only the data table is given to another team. The other team must re-create the site using the data table. When the second mission team has completed its task, the field site should look like the map drawn by the original team.

While monitoring the development of the maps and the re-creation of the sites, the teacher can readily determine which students use proportional reasoning to make their maps, record data involving integers, and use coordinates.

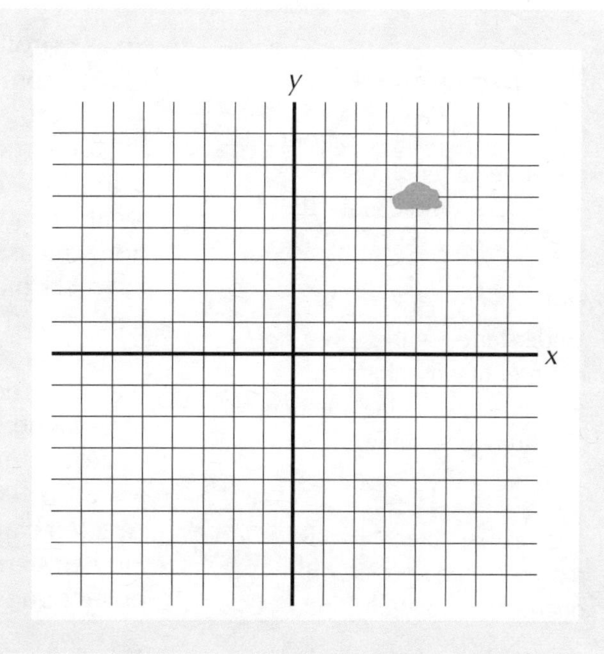

The rock is located at position (4, 5) on this coordinate grid

Concluding the Activity

Each team shows its map to the class and justifies the placement of its rocks on the grid. The locations and justifications should also be turned in as written reports.

On their log sheets, students answer the questions about the mathematics they used and learned in this activity. The teacher should be able to see a correlation among the classwork, class presentations, and students' comments on the log sheets.

ACTIVITY 7

How Big Are the Rocks on Mars?

NCTM Standards

Instructional programs from prekindergarten through grade 12 should enable all students to—

Number and Operations

Understand numbers, ways of representing numbers, relationships among numbers, and number systems

Algebra

Use mathematical models to represent and understand quantitative relationships

Geometry

Analyze characteristics and properties of two- and three-dimensional geometric shapes and develop mathematical arguments about geometric relationships

Measurement

Understand measurable attributes of objects and the units, systems, and processes of measurement

Apply appropriate techniques, tools, and formulas to determine measurements

Astronauts brought back more than 800 pounds of lunar rocks from the Moon between 1969 and 1972. Geologists from around the world have been studying, measuring, and classifying these rocks for the past 30 years. These rocks can be measured using some of the techniques from previous activities, but much of what we have learned about real space rocks is the result of observational measurements. For instance, much of our knowledge about Martian rocks has come from examining pictures transmitted by Mars landers and rovers. Scientists have also learned how to collect data about Mars rocks using rover robotic devices. This scientific advance has provided us with more data than could have been gained from pictures alone.

Rovers must be able to "see" where they are going to navigate on Mars. The IMP, or imager for the Mars *Pathfinder,* used a stereo camera. A stereo camera works in much the same way as human eyes do. Human stereo vision helps people determine the size or distance of objects under observation.

Important Mathematical Ideas

Students use proportional reasoning and angle measures to determine the sizes of objects from a remote location.

Mission

Students simulate the operation of a stereo camera to determine the sizes of, and distances to, objects under study.

Materials and Equipment

A collection of large rocks or boxes; metric rulers; grid paper; copies of protractors; protractors; thin, small-headed pins or paper clips; and log sheets (see appendix, page A-2)

Launching the Activity

Have students try the following two experiments to emphasize the importance of having two eyes separated by a few centimeters rather than just one eye. Have students cover one eye and try to thread a needle. Even big needles with big holes are difficult to thread. Then have students try the same task but with both eyes open. Threading a needle is much easier with the use of both eyes.

The second experiment may be a little messy, but it is worth the cleanup time. Have students close one eye and try pouring water from one container into a second container that has a narrow opening. After the mess is cleaned up, have students try the experiment again with both eyes open.

Discuss each of the experiments. Students should be led to understand that because their eyes are separated, each eye sees a slightly different image. The human brain combines the images into one and uses all the information to determine how close or far away things are, as well as their size. Engineers applied this knowledge as they developed the stereo cameras used on the Mars *Pathfinder.*

Developing the Activity

In this activity, each team will need a workspace that is long and narrow, such as the top of a teacher's desk or a few student desks pushed together.

Give each team a box or other object whose size is somewhere between the size of a shoebox and that of a copy-paper box. If an irregularly shaped object is used, then two points must be marked on the object, one at a location that identifies the right bottom point on the front of the object and one that identifies the left bottom point on the front of the object.

Give each team a metric ruler, a protractor, and two photocopies of protractors. Have students measure and mark off 50 cm along the narrow edge of their work surfaces. This marked-off length will be the baseline for measuring the object. It represents the two "eyes" of the camera. A photocopy of the protractor should be taped to each end of the baseline in such a way that the vertex of the protractor is directly over the endpoint of the baseline.

Students should take turns holding a pin or straightened paper clip at the vertex of the protractor, then kneeling down and "sighting" the corners of the box along the vertex pin. A second pin should be placed along the arc of the protractor to enable students to measure and record the angle that is made between the endpoint of the baseline and the sighting line. This task should be done from both ends of the baseline for both bottom corners on the front of the box. Students should take turns making these measurements, then compare their data as a check for accuracy.

Once the angle data have been collected, students should make scale drawings of the baseline on grid paper. Students may want to use a scale of 1:5, in which the width of one box on the grid paper represents 5 cm along the baseline. Students should be directed to draw the scale model of the baseline close to the bottom edge of their papers to leave room for the rest of the scale drawing.

Have students use protractors to draw the angles measured for the points on both bottom corners of the box. The rays should be extended until they intersect with the corresponding rays from the other endpoint of the baseline. This extension should result in two points of intersection.

A line segment should be drawn between the two points of intersection. The length of this line segment represents the width of the box.

To determine the actual width of the box, students need to construct and solve a propor-

Mars Pathfinder: Roving the Red Planet

Internet Connection

Check out spaceplace.jpl.nasa.gov/urbie_action.htm for a related experiment.

tion. The drawn width of the box is related to the width of the actual box in the same way that the drawn baseline is related to the actual baseline. If the students have used a scale of 1:5, then the same scale applies to the width of the box. One grid box on the scale drawing equals 5 cm. For example, if the width of the box is represented by 3.5 cm, then the actual width of the box is determined using the proportion $1/5 = 3.5/w$, or $1w = (3.5)(5)$; thus, $w = 17.5$, and the width of the actual box is17.5 cm.

Concluding the Activity

Have students check their work by measuring the actual box after determining its width using proportional reasoning. The actual measurements are likely to differ, at least slightly, from the calculated measurements. Initiate a whole-class discussion about reasons for these differences.

This activity prompts students to use mathematics they already know in different ways. Students may also be left with questions regarding mathematics and science in areas related to those addressed in this activity. Have students discuss some of these questions, then record their reflections on their log sheets.

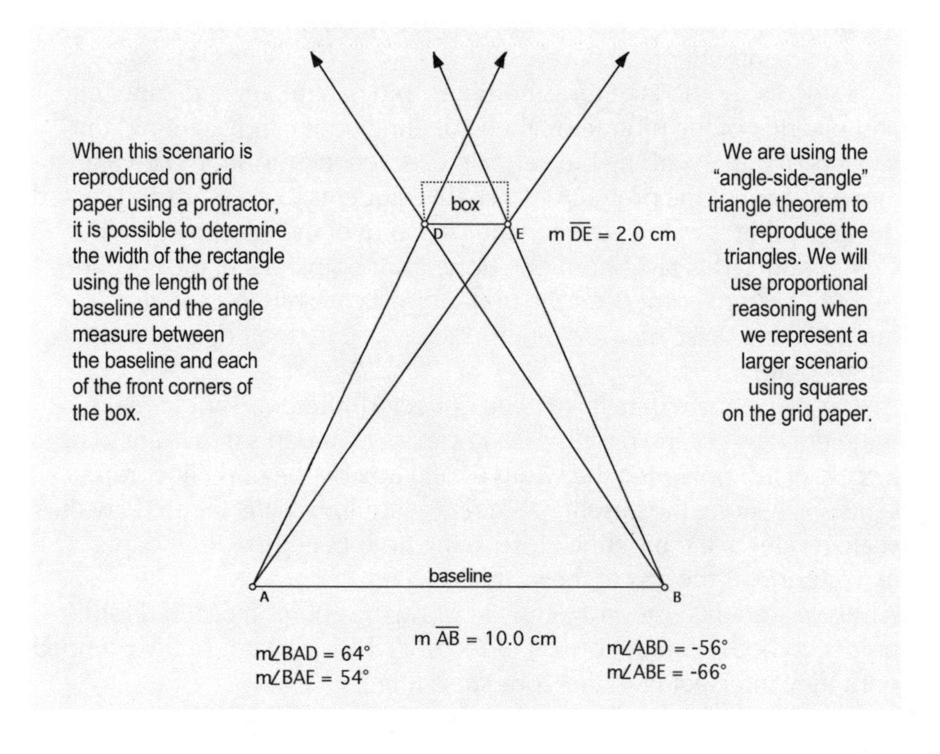

Aeronautics

Aeronautics is the science of flight. Aeronautical engineers solve problems and develop new concepts related to flight. NASA is a major player in this field. In NASA research laboratories and facilities, the many areas of study today include high-speed civil transport, high-performance aircraft, and the effects of airplanes on the environment.

The high-speed civil-transport research, like many other NASA research efforts, is being conducted by a national team. NASA research centers are cooperating with manufacturers, materials suppliers, and universities in research to build an airplane that will carry up to 300 passengers at speeds of around 1600 miles per hour. This plane will not be much faster than the recently retired Anglo-French Concorde, but it will carry more passengers farther because it will be more efficient. The development of the high-speed civil transport is an example of a complex task being performed by teams of experts from many fields.

Middle school students are beginning to think about possible careers. The field of aeronautics should be of interest to students because it shows much promise for employment in the future. The jobs involve more than engineering. Teams of scientists, engineers, technicians, computer scientists, and other support personnel are needed to

An artist's conception of a high-speed civil transport

Internet Connection

The Web site http://nasaexplorers.com/browse_topics.html has lessons for middle school teachers to use in the classroom. Search the site using topics such as "aeronautics."

solve complex problems in aeronautics and to design technology for the next level of development in aeronautics. The fact that these professionals work in teams in these endeavors is significant. Employees in these efforts must be able to cooperate to solve problems and to communicate their results to other teams involved in the task.

In the future, students who learn science and mathematics and have the ability to work in teams may find themselves in careers that address aeronautical problems. Some important questions facing NASA scientists and others today and in the near future include the following:

- How do we keep aging aircraft safe for commercial flight?
- How do we minimize the negative impact of flight on the environment?
- How can the noise of aircraft be reduced?
- What techniques can be developed to predict adverse weather conditions more accurately?
- What kinds of materials can be developed to reduce the weight of aircraft without sacrificing safety?

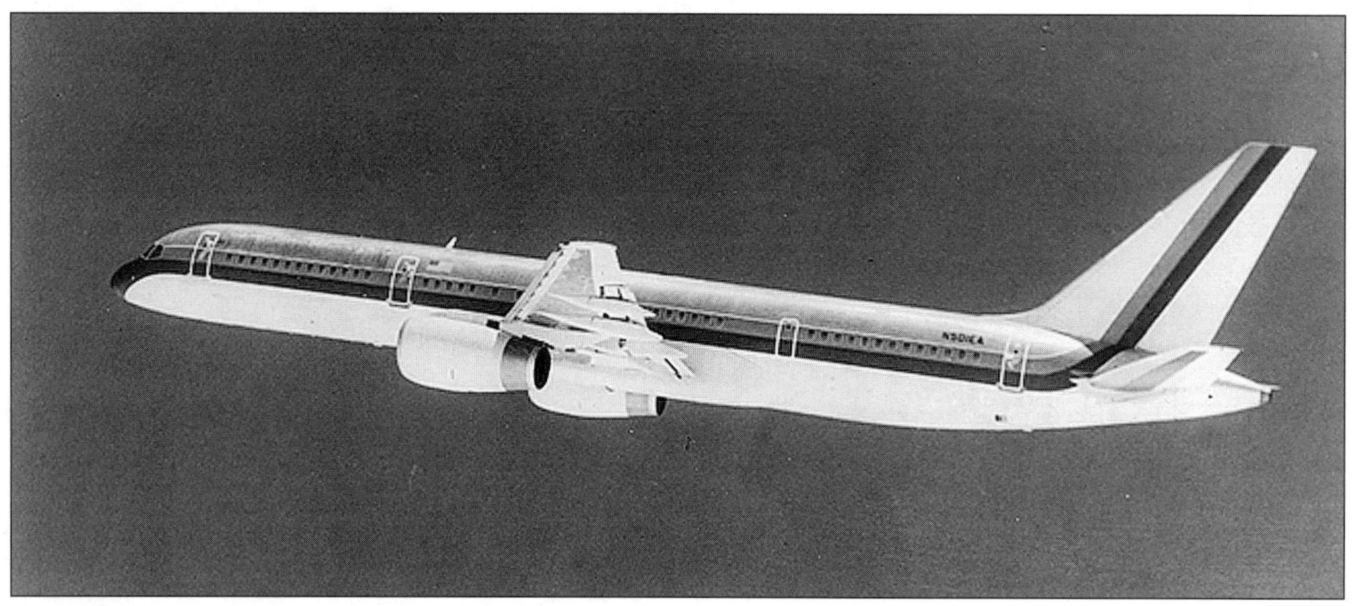

The 757 Flying Laboratory

Air Olympics

The "Air Olympics" module is a set of seven activities leading to a culminating lesson. This module can span thirteen to fifteen class periods, depending on how many optional portions of the activities, supplementary investigations, and assessments are used. The activities need not be done on sequential days. Teachers and students can do one or two activities each week as time allows. The activities are as follows:

- "Designing and Building Paper Airplanes": Students build airplanes of their own design and from patterns.
- "Collecting Test-Flight Data": Students make conjectures about which features of an airplane help it perform well, then take measurements during test flights to test the validity of their conjectures.
- "Organizing and Analyzing Test-Flight Data on Glide Ratios": Students test-fly their airplanes, then collect, present, and analyze data about one aspect of their planes' performance.
- "Collecting, Organizing, and Analyzing New Data": Students identify design features that enhance paper-airplane performance; they then design and build paper airplanes with these features, test-fly the airplanes, and collect and analyze data about test flights.
- "Defining the Events": Students design events that are appropriate for testing different performance factors, such as distance, time aloft, altitude, and all-around best.

A Space Shuttle riding piggyback on a 747

NCTM Teaching Standards

The teachers of mathematics should ... provid[e] and structur[e] the time necessary to explore sound mathematics and grapple with significant ideas and problems.

(NCTM 1991, p. 57)

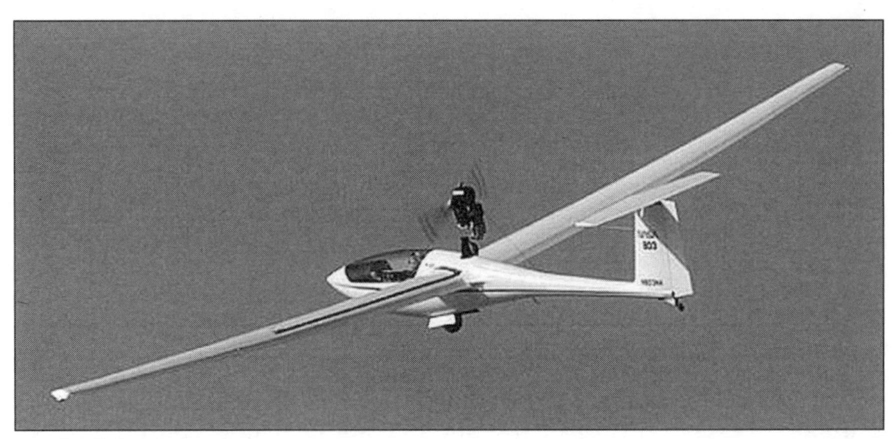
An ultralight airplane

- "Air Olympics": This activity is the culminating event in the module.
- "Debriefing": Students summarize the results of the Air Olympics and revisit their list of conjectures about designing, building, and flying paper airplanes.

The Air Olympics requires advance preparations and clearances from the school administration. Space requirements to fly paper airplanes include the use of a large room, such as the auditorium or gymnasium. An alternative site is a hallway where noise from the activity will not disturb other classes. Given that the Air Olympics promotes active participation from students, parents or other volunteers may be recruited to ensure that students remain on task during the activity.

A Harrier jet

ACTIVITY 1

Designing and Building Paper Airplanes

Introduction

When you first saw—

- a Boeing 747, did you wonder how such a large object could fly?
- a Harrier jet land vertically, did you wonder how it could accomplish that feat?
- an ultralight airplane, did you wonder how such a small plane could fly with a person in it?

Comprehending how these unusual aircraft can fly the way they do may be difficult. The existence of such aircraft comes from the work of aeronautical engineers—the scientists who dream a design, use mathematics and science to design an aircraft, supervise the building of the new plane, and test their designs in controlled flights. In this first activity and the ones that follow, students have an opportunity to simulate the work of aeronautical engineers.

Important Mathematical Ideas

This exploratory activity sets the tone for the remainder of the module. In the process of building paper airplanes, students use the vocabulary of geometry and measurement to describe their work.

Mission

Students use measurement and geometry to design, build, and test-fly paper airplanes.

Materials and Equipment

Each group of students needs the following materials: log sheets (see appendix, page A-2), $8\frac{1}{2}$-by-11-inch paper, a stapler and staples, glue sticks, scissors, cellophane and masking tape, paper clips, plastic drinking straws, rulers, and instructions for the five basic models (see appendix, pages A-9–A-16).

Launching the Activity

In this introductory activity, students begin to learn about making and flying paper airplanes. They should also come to understand that engineers and technicians work in teams to accomplish the development of airplane designs in real life. For this activity, groups of three or four students should be designated as mission design teams, or groups of seven

NCTM Standards

Instructional programs from prekindergarten through grade 12 should enable all students to—

Geometry

Analyze characteristics and properties of two-dimensional geometric shapes

Apply transformations and use symmetry to analyze mathematical situations

Use visualization, spatial reasoning, and geometric modeling to solve problems

Related Mission Mathematics Activities

Mission Mathematics II: Grades 3–5 (Hynes and Hicks 2005) offers additional aeronautics activities involving paper airplanes that may be appropriate for some students.

NCTM Teaching Standards

The teacher of mathematics should pose tasks that are based on—

...

knowledge of the range of ways that diverse students learn mathematics;

and that ...

promote communication about mathematics;

represent mathematics as an ongoing human activity; ...

promote the development of all students' dispositions to do mathematics.

(NCTM 1991, p. 25)

or eight students may be designated as expert teams. Using groups of three or four team members promotes cooperative learning.

Tell students that this set of hands-on, activity-filled investigations culminates in an Air Olympics. Explain that they will build paper airplanes, test the airplanes, and design events for the Air Olympics. This information will help motivate students to do well and will serve as a brief overview of the module.

Begin by showing the class pictures of things that fly, such as kites, hot-air balloons, biplanes, jets, rockets, and the Space Shuttle, to stimulate thinking about how things fly and about the ways that people have modified "flying machines" over the years. Recent newspaper or magazine articles may also offer good ideas to spark discussion. See the handout of a newspaper clipping about a tailless airplane in the appendix (page A-17).

Before building the paper airplanes, be certain to discuss basic ground rules for using airplanes in the classroom and to emphasize the need for safety. Teachers may want to make a large poster or bulletin-board display of these rules for class reference during the activities. The following rules are samples that you may wish to adopt:

- Rule 1: No airplanes may fly in the classroom. All test flights must occur at specified times in designated test-flight areas. Any airplanes flown outside the designated areas or at other than the specified times will be eliminated from the competition, and their pilots will be grounded during the Air Olympics event.
- Rule 2: Safety is our main concern. When airplanes are flown, all personnel will observe safety precautions. Safety goggles will be worn when planes are flying, and only one plane will be in the air at a time.

Students build and test their first paper airplanes using their own designs, then build and test two paper airplanes from the designs provided with this activity. Instructions are included in the appendix for five paper airplanes (pp. A-9–A-16). Many books have been written about paper airplanes, and these often include additional designs. You may want to replace some of the suggested designs or add designs that are familiar to you.

A Blackhawk helicopter

Developing the Activity

Give the following directions to the students:

In your mission design teams, each of you will design and build a paper airplane that you think will fly. Each person will have one piece of paper to make one airplane. As you make the airplane, think about the folds you are using. You should be able to tell someone else how to make the same folds. Try to use mathematical terms, such as triangle, rectangle, pentagon, trapezoid, bisector, *and* symmetry, *to describe your design.*

Have students form pairs with partners from different mission teams. Make sure that students do not show each other their paper airplanes. Have students sit back-to-back, then tell them that they will take turns describing to their partners how to fold their paper airplanes. The catch is that students are not allowed to look at each other's paper airplanes but must rely on spoken design directions. Encourage students to use mathematical terminology in their descriptions. Once students have followed the directions, have them compare their airplanes with the originals. If the designs are different, ask students to determine what might have caused the discrepancy.

Ask for several volunteers to describe how they made their airplanes. Remind them to use mathematical vocabulary to make their descriptions accurate.

Students should return to their design teams to test their airplanes. Ask students to observe test flights of all airplanes in their design teams and to take note of features that seem to make certain airplanes fly better than others. Once each student has flown his or her airplane, have the class make a list of the features that seem to affect the flights. Then have students determine which of these features are qualitative and which are quantitative.

Make a cumulative list of these features on the chalkboard; ask students to name mathematical terms used in describing their airplanes. Reinforce appropriate use of mathematical vocabulary to communicate ideas to others. If a master set of all mathematical vocabulary is developed, the frequency of use of each word can be noted and graphed.

To assess students' understanding of this portion of the activity, have them keep log sheets in their logbooks. Ask students to think about the mathematics they used in making their airplanes. Each student should record directions for making his or her design and the design of one other student in the mission design team. Emphasize the importance of using mathematical terms, such as symmetry, perpendicular, parallel, intersecting, inches, and so on, in the descriptions. Students should also record their observations about the results of their first experience in building planes. These observations could include pictures of successful airplanes. In subsequent activities, students will be asked to be more purposeful in modifying or building planes. Their logbook writings will help them work through these future challenges.

Students will next build a model of the Boeing 757 airplane using a glider kit designed as a part of NASA educational materials. Hand out copies of the glider pattern in the appendix (pages A-9 and A-10). Each student should build this airplane. Discuss the directions, and ask students to identify the mathematical terms used. Ask them to identify points where mathematical terms could have been used in place of other terms. This large-group activity prepares students for building other paper airplanes from patterns.

Teaching and Assessment Tip

By listening to students' discussion during the activity, teachers learn about students' ability to use the vocabulary of geometry and measurement. This knowledge may help in making an informed decision about whether a review of geometry terms is needed before proceeding to the second part of the activity.

Assessment Tip

Logbooks and daily journal writing are important assessment devices built into this set of activities. By reading students' responses, teachers can make better decisions about selecting and developing appropriate mathematics lessons.

Teaching Tip

The work of the expert groups may take more than one class period. In the first class period, students can complete entries in their journals as a "sponge" activity.

A sample logbook entry follows: "I was surprised how much mathematics I used to make my airplane. I have made airplanes before without thinking about mathematics. This class makes me think too much!"

The next part of the activity can be accomplished in larger groups. Students can rearrange their desks to form four "expert groups" for the remainder of the class period.

Each person in the group will build and fly the assigned basic paper-airplane models included in this set of activities.

- Expert Group 1: The Egret[1] (see appendix, pages A-11 and A-12)
- Expert Group 2: The Flex (see appendix, page A-13)
- Expert Group 3: The Basic Square[2] (see appendix, page A-14)
- Expert Group 4: The Condor (see appendix, pages A-15 and A-16)

One goal of this lesson is to use mathematical vocabulary while building paper airplanes. Some of the directions contain few mathematics words. Ask the expert groups to modify and improve the directions using mathematical vocabulary as they build the planes. For example, *line of symmetry* would be more mathematically descriptive than *center line*. Have students record in their logbooks the mathematics vocabulary they added to the directions.

Once the members of each group have built their assigned airplanes and modified the directions, have the expert groups exchange sets of directions. The members of each expert group should work cooperatively to understand the instructions and successfully build their airplanes. Each student should make five airplanes, including the 757.

See appendix page A-17 for information about a unique design for airplanes, a tailless plane. Wingless aircraft are shown below.

[1] The Egret and the Condor are used with permission of Sterling Publishing Co., 387 Park Ave. S., New York, NY 10016, from *Best Ever Paper Airplanes* by Norman Schmidt, ©1994 by Norman Schmidt, a Sterling/Tamos Book.

[2] The Basic Square is used with permission of Workman Publishing Co., from *The World Record Paper Airplane Book* by Ken Blackburn and Jeff Lammers, ©1994 by Ken Blackburn and Jeff Lammers.

Various wingless aircraft—the X-24A, the M2-F3, and the HL-10

Give students an opportunity to fly their airplanes. Instruct them to observe the performance of the airplanes. Which ones fly the best? Students should record in their logbooks the features that seem to make some airplanes fly better than others. These data and those recorded from the first experience with building and flying paper airplanes will be used in subsequent lessons.

Concluding the Activity

Ask students what mathematics words they added to the directions to make them more precise or easier to understand. Lead a large-group discussion in which students share their mathematics vocabulary and explain how the words are used in the directions. Add any new terms to the class vocabulary list. Ask students to show what the terms mean in relation to building paper airplanes.

Teaching Tip

Making mission logbooks with cover designs and mission patches may extend the set of activities for an extra class period or more. Throughout the module, these tasks can become sponge activities to keep students on task while others in the groups complete the activity of the day.

ACTIVITY 2

Collecting Test-Flight Data

NCTM Standards

Instructional programs from prekindergarten through grade 12 should enable all students to—

Measurement

Understand measurable attributes of objects and the units, systems, and processes of measurement

Apply appropriate techniques, tools, and formulas to determine measurements

Data Analysis and Probability

Formulate questions that can be addressed with data and collect, organize, and display relevant data to answer them

Develop and evaluate inferences and predictions that are based on data

In activity 1, students had an opportunity to explore making paper airplanes and using geometry vocabulary. "Collecting Test-Flight Data" requires students to reflect on their experiences in the first activity to gather more purposeful data about the flight of paper airplanes. In the real world of aeronautical design and testing, having a good idea may not be sufficient to build a plane. An engineer must be willing to state that idea to others on the team. The idea must become public. Then the other team members demand data to verify the idea. When the idea seems feasible, the whole design team becomes involved in collecting more data to convince others outside the team that the idea is valid. Measurement is often used as a tool to collect these data.

Important Mathematical Ideas

The purpose of this activity is to have students collect data about the design and performance of the paper airplanes that they have built. Although students should be encouraged to continue to use geometry vocabulary to describe their designs, measurement becomes a new focus in this lesson. Students must learn to describe where they have made a fold or a cut when building a paper airplane. They also need to use measurement to record data about the performance of their airplanes.

Mission

Students will identify design features that enhance paper-airplane performance and will collect data during test flights to test conjectures about the design of paper airplanes.

Materials and Equipment

Log sheets (see appendix, page A-2), including a list of conjectures from the large-group discussion conducted in activity 1; airplanes from activity 1; paper; tape; and scissors. Students may request stopwatches, metersticks, and measuring tapes.

Launching the Activity

The initial part of the activity should be done as a whole class. In a teacher-led discussion, students reflect on the performance of their first paper airplanes, both their own designs and the model designs. Make sure that the discussion covers the following questions:

- What features of the airplanes made them perform well?
- What construction techniques seemed to make a particular airplane a good one?

Students should be encouraged to offer their conjectures about these and other questions. Remember, when students offer their thoughts, they are exposing themselves to criticism from their peers. Setting the proper tone for the class is important to avoid embarrassing students. Successful teams are those that focus on ideas, not on the person making the suggestions. Even those conjectures that are eventually rejected are important to the development of new knowledge about the problem. In real life, some conjectures that seemed outrageous at first have led to significant discoveries, even to Nobel Prizes.

All new conjectures should be recorded on the chalkboard for the whole class and in individual logbooks by each student for future reference.

Developing the Activity

After the large-group discussion, students should be given the following tasks to complete in their four-member design teams:

- Use measurement to collect and record data about "good" airplanes.
- Use the airplanes from activity 1 to test conjectures from the class discussion.
- Select an airplane thought to be "a good one," and use measurements and geometry vocabulary to describe how to build the plane.

A Drone for Aeroelastic Structure Testing (DAST)

NCTM Teaching Standards

The teacher of mathematics should promote classroom discourse in which students—

...

make conjectures and ...

try to convince themselves and one another of the validity of [their] conjectures....

(NCTM 1991, p. 45)

Remind students about the safety rules for this module. Also, inform the teams that at the conclusion of the data collection, one team member will report the results of their work. Teams should select this representative before collecting their data.

Each design team begins by selecting conjectures and airplanes to help them test the conjectures. Next, members of the team get clearance to go to the data-collection area, where they can fly airplanes and take test-flight measurements. The teams must also report airplane measurements in addition to their test-flight data.

No direction has been given about what type of measurements to make during the test flights. Some students may measure the distance of the flight, whereas others may count the number of turns an airplane makes before landing. This variety in data collection should be encouraged.

Teachers may want to limit the number of test flights to avoid confusion in the classroom and to keep within the class time limit.

After collecting data, the design teams confer about their observations and results. The discussion should end with each student on the team writing an evaluation of the conjectures he or she tested and the airplanes used to collect the data.

Concluding the Activity

Bring the class back together to conclude the activity. The selected member from each design team should report the results of the team's work. Make sure that students understand that they are to learn from these reports, not critique them. Test results from one group may give insight into a conjecture or airplane design that another group had forgotten to consider.

Record which conjectures were tested and what types of airplanes were used to test the conjectures. Encourage students to describe their airplanes using measurements. Many airplanes have similar shapes but different sizes or designs.

Encourage students to record notes in their logbooks during the presentation of results.

At the conclusion of the large-group discussion, instruct students to return to their design teams to consider the data collected by all the groups. After this small-group discussion, the members of each group are asked to write in their logbooks three conjectures regarding the characteristics of a paper airplane that make it fly well. These conjectures should be based on the data the group members have seen and collected themselves. The students in each design team then discuss the new ideas and agree on three new conjectures to report to the whole class at the beginning of the next activity.

ACTIVITY 3

Organizing and Analyzing Test-Flight Data on Glide Ratios

Flight-testing experimental aircraft is a serious scientific endeavor. Aeronautical engineers at NASA design many kinds of tests to ensure that airplanes perform consistently and safely. Engineers test the controls, stress on the wings and rudders, fuel consumption, cockpit instruments, and many other aspects of the planes during test flights. They also test each plane's overall performance, that is, how far it can fly on a full load of fuel, how sharply it can turn or dive, how much runway it needs to take off and land, and whether its onboard computers are capable of controlling the airplane under different circumstances. To develop tests that will yield meaningful information, scientists must determine what data they will collect, how the data will be organized, and which statistical measures will be used in analyzing the data.

Important Mathematical Ideas

The purpose of this activity is for each team to conduct a flight test and analyze, compare, and discuss the resulting data as a class. Through this activity, students should learn that extraneous variables must be controlled when testing other variables and that careful measuring and recording of data are imperative in making meaningful conclusions.

Mission

Students test-fly their paper airplanes and collect, present, and analyze data about one aspect of the planes' performance.

Materials and Equipment

Log sheets (see appendix, page A-2), paper airplanes, stopwatches, metersticks, measuring tapes, graph paper, and an overhead projector. Students may also want to use a camcorder to record flights. A test-flight data sheet is provided in the appendix on page A-18.

Launching the Activity

In the previous activity, students made a list of conjectures regarding the characteristics of airplanes that make the airplanes fly well. This list should be displayed in the classroom. After flying their airplanes, students recorded three new conjectures in their logbooks.

As a class, review the displayed list of conjectures and have each group report three new or refined conjectures. Revise the first class list

NCTM Standards

Instructional programs from prekindergarten through grade 12 should enable all students to—

Algebra

Use mathematical models to represent and understand quantitative relationships

Measurement

Understand measurable attributes of objects and the units, systems, and processes of measurement

Apply appropriate techniques, tools, and formulas to determine measurements

Data Analysis and Probability

Formulate questions that can be addressed with data and collect, organize, and display relevant data to answer them

Select and use appropriate statistical methods to analyze data

Develop and evaluate inferences and predictions that are based on data

Test-Flight Data for Airplane #1

Number of Steps

9 8 7 6 5 4 3 2 1 0

Flight 1 Flight 2 Flight 3 Flight 4

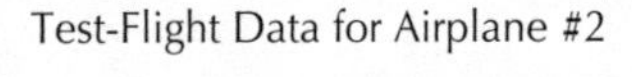

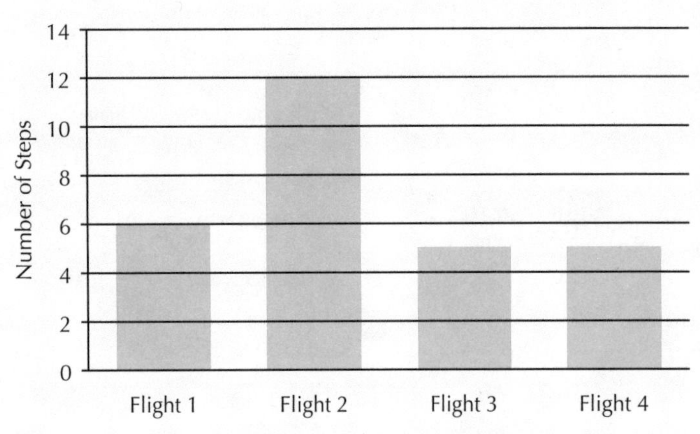

Test-Flight Data for Airplane #3

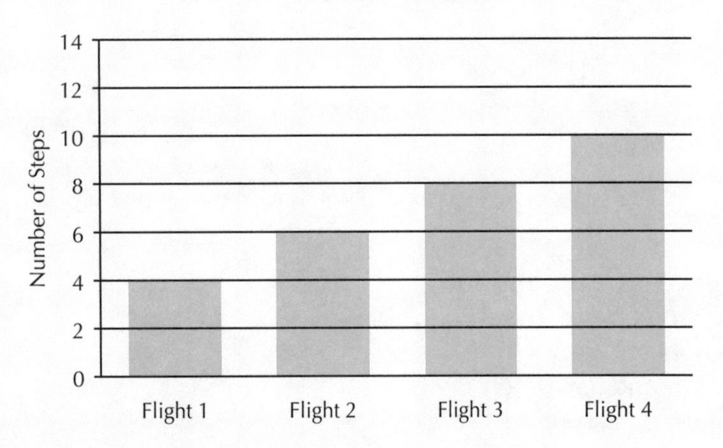

in the course of this discussion. Some conjectures will be supported and some, rejected. Some preliminary conjectures may be revised or refined, and some new conjectures may be proposed. The spirit of the discussion should focus on sharing ideas and thinking about collecting more data to test the revised list of conjectures.

Before students reassemble in their design teams, discuss the organization and analysis of data by sharing an example of the process with the whole class. The following example should stimulate students' thoughts about collecting, organizing, and analyzing different types of data.

> *In a similar investigation to the one we are conducting today, one of the design teams collected some data that I thought I would share with you. The members of this group were impressed by how high their airplanes could fly. They wanted to collect data about this aspect of their airplanes. They decided to use the stairway in the school hallway to record the data. They stood at the base of a staircase and tossed their airplanes toward the top step. The students recorded the number of the step where each airplane landed. They repeated this process four times with three different paper-airplane designs. The graphs at the left show the data they collected and the way they organized their data.*

Continue using this example to discuss how the data can be analyzed. Point out that the labels on the graphs are important in determining what the graphs represent. Tell the students that on the basis of these data, the design team was trying to decide which airplane would be the best for the Air Olympics "stair-throw" event, which tests how high an airplane can fly. The depth of this discussion regarding data analysis will depend on the students' experience with statistics. Lead the discussion with such questions as the following:

- What makes you think this team collected accurate data?
- How did the students organize the data in a meaningful way?
- How would you analyze these data?
- Which airplane would you select to fly in the stair-throw event?

If students have had experiences with mean, median, and mode, tell them that because of the results from airplane 2, this design team chose not to use the mean to evaluate its data. Ask students to justify the design team's choice on the basis of these data.

If data from all teams are considered, other data displays may be used. Stem-and-leaf plots would be suitable if the staircase has thirty or forty steps. A box plot would help students identify paper airplanes that perform above and below reasonable expectations. The data for this example are depicted in bar graphs. Students may have studied other types of graphs and plots to portray data. If students have not had experience with statistics, interspersing statistics lessons with this module may be appropriate. Students might then revisit data analysis for this activity.

Students should record in their logbooks the name and description of each type of data display used. They might also show examples of each type of data display.

Developing the Activity

Begin the next part of the investigation by asking the following questions: What would happen to an airplane if it suddenly lost its power? Would it drop like a rock? Would it glide for a while? How far could the pilot expect to fly it before reaching the ground?

Tell students that the *glide ratio* of a plane helps to answer questions like these. It describes the horizontal distance that the plane will travel from a given altitude in the absence of power and wind. The glide ratio is usually expressed as a negative number because the change in altitude is a loss of altitude and is, therefore, expressed as a negative distance. Students might benefit from a discussion of the fact that temperatures below zero are represented by negative numbers, as well.

Have the design teams choose one of their paper airplanes that they think will stay aloft and glide a long distance, even with a light toss, and one airplane that they think will fall quickly.

Take the class to a previously set-up area to conduct test flights. This area should have a starting line that is wide enough for a member of each team to launch an airplane without the airplane colliding with that of another team. Each team should also have a measuring tape to find the distances its planes travel. Have students collect data from four launches.

Concluding the Activity

Gather students as a class to compare and discuss the test flights and data. Guide students to the conclusion that testing situations must be consistent. Students may suggest that airplanes need to be launched with a toss of similar force or from the same starting point. If students do not make this suggestion, ask probing questions to lead them to this conclusion.

Students will likely suggest that each launch should be made from the same height. This suggestion presents the opportunity to discuss how the glide ratio is determined. The glide ratio is determined by dividing the horizontal distance flown by the change in altitude. This ratio describes the average slope of the airplane as it descends; it should be the same regardless of the height of the launch. Conducting a mini-investigation of this concept may be worthwhile if students seem puzzled by this discussion.

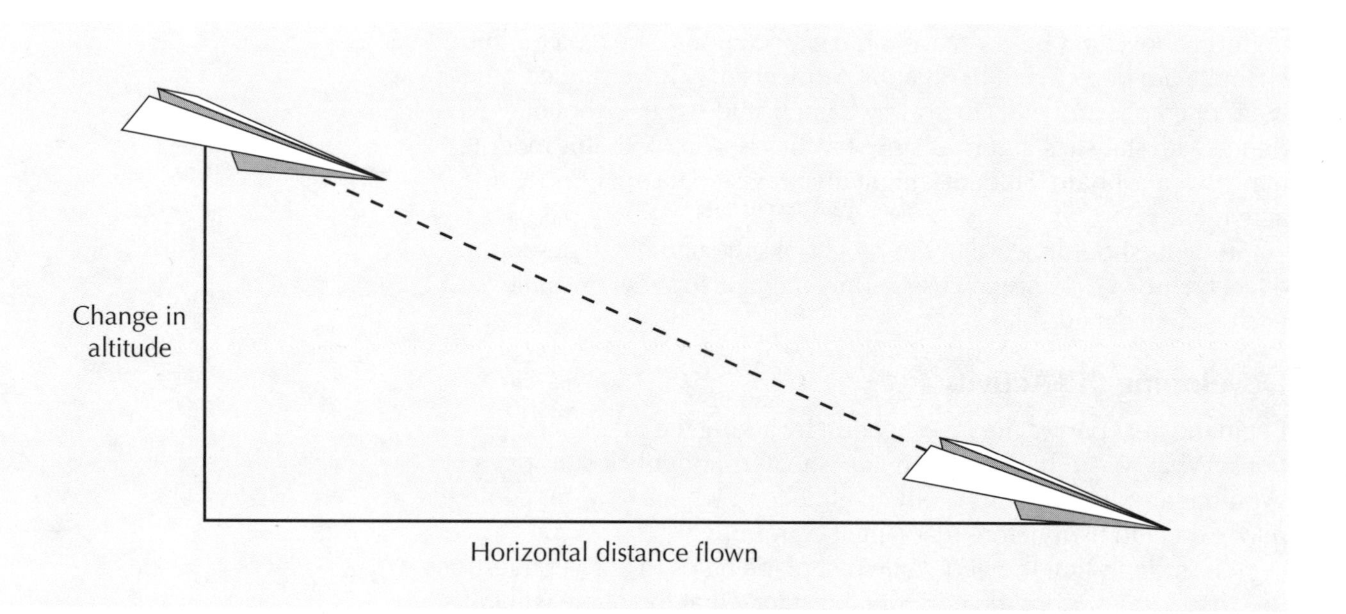

After students agree on ways to control the force of the launch, return to the test area and collect data on ten flights. Remind students to collect data on both the launch height and the distance flown.

Have each team calculate the glide ratio for each launch. Students should have several different glide ratios with which to work. Lead a class discussion on how to manage the data. Analyze, compare, and discuss these data as a class.

Students should record on their log sheets what they learned from this experience.

Extending the Activity

Depending on students' experiences with measures of central tendency, this activity may offer an opportunity to discuss which measure would be most appropriate to summarize the data—the mean, the median, or the mode.

ACTIVITY 4

Collecting, Organizing, and Analyzing New Data

In activity 2, students worked in their design teams to investigate characteristics of their airplanes that might make them fly "better." The activity emphasized the hypotheses that could be made and investigated regarding characteristics of paper airplanes rather than the ways in which data were collected, displayed, and analyzed. In this activity, the design teams revisit their refined conjectures as they work to develop a test plan for more structured investigation of their hypotheses.

Important Mathematical Ideas

In this activity, students fly their airplanes and collect data about airplane performance during the flights. The test-flight data sheet on page A-18 is designed for students to use line plots to record and analyze their data. Other organizing devices, such as stem-and-leaf plots, scatterplots, and various graphs, may require further explanation in class.

Mission

Students identify design features that enhance paper-airplane performance. They then design and build paper airplanes with these features, test-fly the airplanes, and collect and analyze data about test flights.

Materials and Equipment

Log sheets (see appendix, page A-2), test-flight data sheets (see appendix, page A-18), paper airplanes, stopwatches, metersticks, measuring tapes, graph paper, and an overhead projector. Students may also want to use a camcorder to record their test flights.

An X-3

Launching the Activity

Direct students to return to their design teams and collect more data about their airplanes. Each design team will focus on one thing that their airplanes seem to do well.

NCTM Standards

Instructional programs from prekindergarten through grade 12 should enable all students to—

Measurement

Understand measurable attributes of objects and the units, systems, and processes of measurement

Apply appropriate techniques, tools, and formulas to determine measurements

Data Analysis and Probability

Formulate questions that can be addressed with data and collect, organize, and display relevant data to answer them

Select and use appropriate statistical methods to analyze data

Develop and evaluate inferences and predictions that are based on data

NCTM Teaching Standards

The teacher of mathematics should pose tasks that are based on—

...

knowledge of the range of ways that a diverse group of students learn mathematics;

and that—

...

promote communication about mathematics;

represent mathematics as an ongoing human activity;

promote the development of all students' dispositions to do mathematics.

(NCTM 1991, p. 25)

Teaching Tip

Encourage students to use calculators for this part of the activity. Calculators are tools that can be helpful in making decisions about how to score an event.

For example, their planes may fly a long distance, they may stay aloft a long time, or they may make many loops and turns before they land. Their airplanes may fly high, like those used to collect the sample data.

Tell the class that each design team will be responsible for constructing a test plan. This test plan must be submitted for teacher approval before conducting test flights. The test plan should include the aspect of the airplane to be tested, the manner in which data will be collected, the type of data to be collected, and the planned analysis of the data.

Once the test plans have been approved, the design teams should move to the data-collection area to continue the investigation. Data from each design team will be recorded on the test-flight data sheet.

Developing the Activity

This activity focuses on the mathematical task of analyzing data. Once students have collected their data, check to be sure that they are conducting this analysis. Make a checklist for each team to ensure that its data are organized and represented appropriately using graphs. Next ask the team members what measures of central tendency they will use to analyze their data. They may choose to use the mean, median, or mode, depending on their previous experiences with statistics and on the nature of the data they collected. Be sure to suggest that each team justify its choices.

Once the students in each design team have analyzed their test-flight data, they should reflect on their results and make conjectures about which design features enhance the aspect of performance that they tested. Students then modify their planes or design new ones and test-fly the new designs to test their conjectures.

After completing the test flights, students should record the results of their efforts in their logbooks. One member of each team should record the results on an overhead transparency to share with the class in the closing activity. Each team should also display its results graphically.

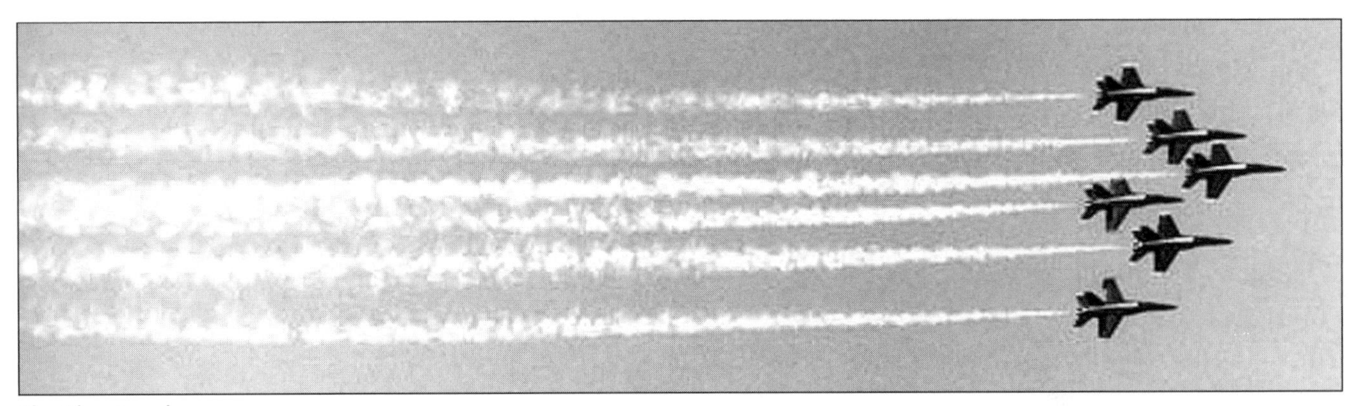

The Blue Angels

Concluding the Activity

Each design team should report to the whole class about its test flights, sharing the data as organized by the group. Class discussion should focus on using the vocabulary and concepts of statistics. Design teams should be prepared to support their conclusions on the basis of the data they collected and the choices they made for analysis.

Lead the discussion back to the list of class conjectures about paper-airplane flight. After all design teams have reported, students should think about whether the data collected in the test flights confirmed, altered, or eliminated any of the class conjectures. If new conjectures are developed, they should be added to the class list.

An X-3

NCTM Teaching Standards

The teacher of mathematics should promote classroom discourse in which students ... try to convince themselves and one another of the validity of particular representations, solutions, conjectures, and answers....

(NCTM 1991, p. 45)

Assessment Tip

Assessment practices should shift toward integrating assessment with instructional activities. By integrating statistics into this activity, the teacher can make instructional decisions that are based on the understanding of statistics shown by students.

ACTIVITY 5

Defining the Events

NCTM Standards

Instructional programs from prekindergarten through grade 12 should enable all students to—

Measurement

Understand measurable attributes of objects and the units, systems, and processes of measurement

Apply appropriate techniques, tools, and formulas to determine measurements

Data Analysis and Probability

Select and use appropriate statistical methods to analyze data

Develop and evaluate inferences and predictions that are based on data

Teaching Tip

Two days may be required to complete the rotations through the events. All students should have the experience of both flying airplanes and judging events.

This activity involves the development of events for the Air Olympics. The purpose of the Air Olympics is to allow competitive testing of airplane designs. Although no parallel competitive situation exists in the world of commercial aviation, engineers continually create test situations for their new designs. Test pilots fly newly designed airplanes according to specific flight plans to ensure that particular aspects of the airplanes are tested. Engineers decide what aspects of the plane need to be tested, develop tests, and identify methods for collecting data during the test.

Important Mathematical Ideas

Students simulate the planning process of aeronautical engineers by creating and refining their paper-airplane designs in an Air Olympics. In this lesson, students devise the events to be included in the Air Olympics. This process involves identifying important capabilities of paper airplanes and determining how to "score" the performance of the airplanes as they compete in each event.

Mission

Students create events for the Air Olympics and develop data-collection techniques and scoring schemes for each event.

Materials and Equipment

Log sheets (see appendix, page A-2), an event specification form for each expert group, paper airplanes, tape, string, stopwatches, metersticks, and measuring tapes

Launching the Activity

In this activity, students conduct test flights to help define the events of the Air Olympics. Students may need to use space outside the classroom to try out ideas about the design of events.

Tell students that their first task is to determine the important features of paper airplanes. Lead a discussion by asking questions about what makes a good paper airplane in terms of performance. Remind students of their previous experiences with flying paper airplanes. Students will probably identify the following capabilities of their paper airplanes as being important:

- Some fly farther than others.
- Some stay aloft longer than others.
- Some can be controlled by the "thrower."

- Some seem to do tricks, such as loops and turns, during flight.
- Some fly higher than others.

Students are usually quite creative when devising events to showcase these capabilities. The class should conduct a brainstorming session to come up with a list of events, then pare down the list to four events through a consensus-building discussion.

Developing the Activity

At this point in the activity, students should return to their four expert groups from activity 1. Each group will focus on the details of a single event, considering how to collect data for the event and designing a scoring scheme.

Remind the expert groups of scoring schemes used in such other events as diving, gymnastics, track, and swimming contests. Many judges may score such events, and often, such statistics as mean, median, and mode are used to summarize the judges' scores.

The challenge to each expert group is to design a good scoring scheme for a single event. Distribute to each student a copy of the Air Olympics event specification form on page A-19. Have each group discuss the various aspects of the form and come to an agreement on each part. In completing the form, students may need to conduct additional test flights, collect additional data, and test their scoring schemes.

Concluding the Activity

After students in each group have come to a consensus regarding the event specification form, have the groups come together as a class. Have one member from each group describe the group's event and the scoring system to be used for the event. These presentations will initiate discussion of the statistics used in the scoring systems.

Often, students describe a scoring system without using statistics vocabulary. Use this opportunity to reinforce statistics vocabulary and concepts. For example, students may describe a three-judge scoring scheme that uses the middle score of the judges as the event score. This scheme could prompt a review of the concept of median and related vocabulary.

Students should record in their logbooks their groups' final agreements about event descriptions, rules, and scoring; they should record descriptions of their newly acquired understandings in mathematics on their log sheets. Encourage students to use mathematical vocabulary in these descriptions.

A wake vortex study

Teaching Tip

Students compute scores for each participant in each event. The accuracy of these computations using the scoring schemes devised in previous activities is important to each pilot. Calculators should be used to perform the initial calculations and to check written computation.

Assessment Tip

Openness is one of the Assessment Standards. In this module, students are aware of the culminating event, the Air Olympics, from the first activity and are given the opportunity to learn the expectations for the events and the methods of scoring. This type of openness should be used in all instruction.

ACTIVITY 6

Air Olympics

NCTM Standards

Instructional programs from prekindergarten through grade 12 should enable all students to—

Data Analysis and Probability

Select and use appropriate statistical methods to analyze data

Develop and evaluate inferences and predictions that are based on data

At test-flying time, aeronautical engineers turn the testing over to pilots. Final data about the performance of the test airplanes are collected while engineers sit on the sidelines and wait to conduct their analysis.

In this activity, students act in several capacities. For example, they act as engineers, with a personal interest in the performance of their designs. Further, students play the role of test pilots as they fly their airplanes. Finally, students measure the performance of airplanes during the test flights of other groups.

This activity serves as the culmination of the module. The events of the Air Olympics may be photographed or videotaped to document the learning that has occurred in this module.

Important Mathematical Ideas

Students simulate the testing process of aeronautical engineers with their paper-airplane designs in an Air Olympics. In this lesson, students take measurements while performing the events planned in activity 5 and conduct data analysis to score airplane performance in the events.

Mission

Students fly paper airplanes in the Air Olympics events and judge the performance of other students' airplanes.

Wilbur and Orville Wright experimented with gliders before building their first powered flying machine in 1903. That machine is shown above after it was damaged in the first trial of December 14, 1903. Three days later, Orville Wright flew the repaired machine on its historic first flight of 120 feet in 12 seconds. (Photo courtesy of the Library of Congress)

Materials and Equipment

Logbooks, paper airplanes, tape, string, stopwatches, metersticks, measuring tapes, cameras, and camcorders

Launching the Activity

Divide students from the four expert groups into two new groups, X group and Y group. One group participates in the events while the other supplies the judges and officials for the events, as follows:

- Rotation 1: X-group members are the judges, and Y-group members are the paper-airplane pilots.

- Rotation 2: Y-group members are the judges, and X-group members are the paper-airplane pilots.

Randomly assign students in the first pilot group to one of four flight squadrons. These flight squadrons rotate through the events using the following scheme:

- Flight Squadron 1: Event 1, Event 2, Event 3, Event 4
- Flight Squadron 2: Event 2, Event 3, Event 4, Event 1
- Flight Squadron 3: Event 3, Event 4, Event 1, Event 2
- Flight Squadron 4: Event 4, Event 1, Event 2, Event 3

The assignment of students to the judging group should be more purposeful. Each of the four judging teams should include students who are familiar with the design of particular events.

Developing the Activity

Take students to the event areas, and begin the Air Olympics. Parent and staff volunteers should oversee the safety and operational procedures of each event. The teacher serves as the overall marshal for the Air Olympics. The role of the marshal is to resolve disputes and answer questions about interpreting the rules.

When rotation 1 is complete, students return to a whole group. The students are then reassigned according to rotation 2, and the Air Olympics continues.

NCTM Teaching Standards

The teacher of mathematics should create a learning environment that fosters the development of each student's mathematical power by providing a context that encourages the development of mathematical skill and proficiency ... and by consistently expecting and encouraging students to ... display a sense of mathematical competence by validating and supporting ideas with mathematical argument.

(NCTM 1991, p. 57)

Concluding the Activity

When all students have completed all four events, they should record their Air Olympics scores in their logbooks and return to the large group. In the next activity, the class data can be summarized by design teams from the building and testing activities.

Summarizing team data gives the class another chance to discuss mean, median, and mode. Because the median tends to compensate for extreme data, this central-tendency measure should be used to summarize team scores. Median scores often make team scores in each event close, adding excitement to the activity. Ask students to begin thinking about how they can compare the results of team scores.

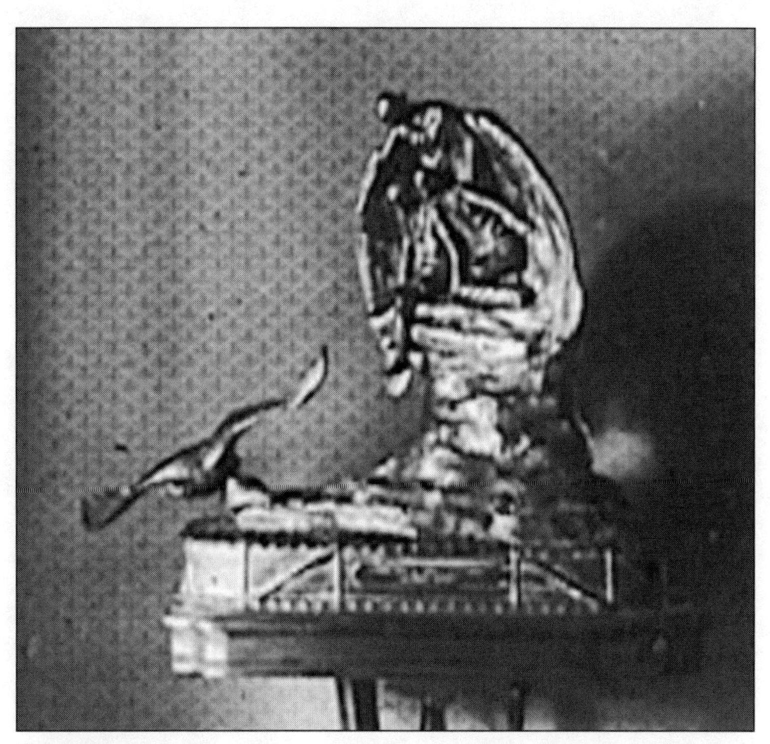

The Muse of Aviation trophy was given to the Wright brothers by the Aéro-Club de la Sarthe in Le Mans, France, in 1909. (Photo courtesy of the Library of Congress)

ACTIVITY 7

Debriefing

NCTM Standards

Instructional programs from prekindergarten through grade 12 should enable all students to—

Data Analysis and Probability

Select and use appropriate statistical methods to analyze data

Develop and evaluate inferences and predictions that are based on data

The debriefing is the final activity of this module. Aeronautical engineers view this step as essential in the process of developing new airplanes. The data from the first cycle of designing, building, and test-flying new aircraft are used to begin a new cycle of development. Experimental designs may require many cycles of redesign and testing. The process of developing better aircraft continues even after experimental designs reach production. Even as production aircraft carry people and cargo, engineers from the aircraft companies collect data to improve the performance of future airplanes.

In this activity, the winners of the Air Olympics events are recognized. Awards, certificates, and thanks are given to all participants. Because the presentation of awards is time-consuming, more than one class period may be required for this activity.

Important Mathematical Ideas

In this culminating activity, students summarize their team data and, for the final time, review the class list of conjectures about paper airplanes.

Mission

Students share summaries of results from the Air Olympics and complete the list of conjectures about designing, building, and flying paper airplanes.

Materials and Equipment

Logbooks and poster board

Activity

Grouped into their design teams, students discuss how to determine team scores.

Because each design team has four members, four individual scores are available to determine the team score for each event. The following sample questions should be considered to summarize the data for the events:

- Should students use the mean of the individual scores for the team score?
- Should students use the median of the four scores? (The median of an even number of data points is the mean of the middle two data points.)

Once team-scoring agreements are reached for all four events, students record their team results in their individual logbooks and on poster board. The poster-board display should include outstanding individual results and the summarized team results.

Developing the Activity

As a class, review the conjectures that were recorded during the module. Discuss each conjecture in light of the individual and team results of the Air Olympics.

As a final assessment activity, each student should prepare a list of conjectures for the module. Some of the conjectures will be those from the class list. Some will be modifications of those on the list as a result of the class discussion. Some conjectures will be new because new learning has occurred during the "Air Olympics" module.

Students should be reminded that this activity models how progress in aeronautical engineering occurs. A continual review of the principles of flight using new materials and technology allows engineers to build bigger, safer, faster, more ecologically sound, and more comfortable airplanes.

Concluding the Activity

To complete this module, recognize the "winners" of events in the Air Olympics. Of course, any awards ceremony should be consistent with school policies. Although the time used to make the awards does not contribute directly to learning, the motivation for learning that is generated through recognition is important for all students. Perhaps all students who compete in the Air Olympics should receive some recognition for their efforts. Students may also wish to represent their experiences in the "Air Olympics" in the design of mission patches or logbook covers.

World-record-breaking paper airplane inside a hangar at Langley Research Center.

Airplanes and Airports

Internet Connection

An interesting Web reference on airports is "NASA Future Flight Central," found on this site: ffc.arc.nasa.gov/index.html

The previous lessons involved the design and testing of paper airplanes. If the testing areas were outside, students may have noticed that the wind affected the flights of their airplanes. Students may also have experienced some paper-airplane collisions during data collection. For paper airplanes, the consequences of such collisions might be only the need to make an additional toss to collect data or, at most, the need to straighten a bent airplane nose before flying the airplane again. The consequences to real airplanes are much more serious, as is the effect of the wind on takeoff and landing. Airports are designed to minimize the risk and maximize the efficiency of flying. NASA FutureFlight Central is a full-scale research-and-development simulator of airport operations that looks and feels like an actual air traffic control tower. This facility conducts research in building and operating more cost-effective and safer airports.

"Airplanes and Airports" consists of a set of five activities designed to help students learn about how runways are named, as well as how pilots and air traffic controllers determine which runway a pilot should use for taking off and landing. Students use the compass rose and measuring tools to determine angles; they look for patterns in data to determine rules for naming runways; and they use problem solving and measurement to determine a flight plan. The activities are as follows:

- "Measuring Angles with the Compass Rose": Students relate angle measures to the compass rose.
- "Exploring the Meaning of Runway Names": Students apply what they know about angle measures to determine the method used to name airport runways.
- "Which Way Should I Land?": Students explore the effects of strong winds from particular directions on the ability of airplanes to land at specific airports.
- "Making and Using a Windsock": Students design and construct windsocks and use them to determine the prevailing wind direction at their school.
- "Designing Runways for Your School": Students use the data collected on prevailing winds to plan an airport for their locality.

NASA Ames Research Center's FutureFlight Central, a fully interactive air traffic control tower simulator.

ACTIVITY 1

Measuring Angles with a Compass Rose

NCTM Standards

Instructional programs from prekindergarten through grade 12 should enable all students to—

Geometry

Specify locations and describe spatial relationships using coordinate geometry and other representational systems

Measurement

Apply appropriate techniques, tools, and formulas to determine measurements

Pilots and air traffic controllers use the compass rose to determine compass headings for flight plans, which are the routes pilots will fly, and to describe the direction pilots are flying. The measures are determined using degrees on a circle; 0 degrees is at the north point of the circle, as it would be if the circle were drawn on a map, and 180 degrees is at the south point. East is at 90 degrees, which is different from what most students and teachers are accustomed to when measuring angles in school settings. In school settings, angles are measured in the counterclockwise direction. When using the compass rose, angles are measured in the clockwise direction.

Important Mathematical Ideas

This activity introduces students to the compass rose and makes a connection between angle measures and magnetic headings.

Mission

Students construct a compass rose using a compass, a protractor, and a straightedge.

Materials and Equipment

Compass, protractor, straightedge, paper, and log sheets (see appendix, page A-2)

Launching the Activity

Give each student a compass and a protractor, and have students use the compass to make a circle on a blank piece of paper. The procedures for using a compass may need to be taught or reviewed, depending on students' previous experiences. Students may make the radius of the circle too large or place the center of the circle too close to the edge to allow them to draw the full circle on the paper.

Each student should be sure that the center of the circle is visible. On the basis of the type of protractor used, students may need to redraw the center of the circle to ensure that it can be seen easily. Have students draw a point on the circle toward the top of their papers and label the point N.

Tell students that the point they have just drawn represents the direction north. Have students use the straight edge of their protractors or a ruler to draw a line segment connecting the center of the circle to point N on the circle. This line segment will be used later when students draw angles on this compass rose.

Ask students where they would draw point S to represent the direction south. Depending on the students' map skills, showing a map during this exercise may be helpful. Once students have identified south as being at the opposite point on the circle, have them use the straight edge of their protractors or a ruler to extend the line segment to intersect the opposite point on the circle. Students should label this point S.

Remind students that a circle is 360 degrees. Tell students that point N is at 0 degrees. Ask students to guess what degree measure would describe the location of point S. What would the degree measures be for east and west?

Have students work in pairs to determine each of these degree measures and to mark the locations for east and west on their circles, labeling the points E and W, respectively. Be sure to check that students do not confuse the directions for east and west as they are working.

Developing the Activity

Once students have identified and labeled points for north, south, east, and west, discuss other direction names that are frequently used in information about wind direction. For example, students may have heard information about wind direction given on a television weather channel or in local weather forecasts. Students will probably suggest such directions as northwest, southeast, and so on. Ask students why "northsouth" is never given as a direction. If students do not come up with an appropriate answer on their own, ask questions prompting them to think about the confusion that might be caused if an observer did not know how east or west relates to that description.

Use a presentation-size compass and protractor or an overhead-projector compass and protractor to model how to identify the location of northwest on a compass rose. Draw a 270 degree angle with the vertex at the center of the circle and the endpoints on the north and northwest locations. Students may need to be shown how to use a protractor to measure angles on a circle. Be sure to use the angle measures on the protractor that start at 0 on north and to measure in the clockwise direction. Direct students to work in pairs to identify the degree measures and draw angles for northeast, southeast, southwest, and northwest on their own compass roses.

Another tool that can be used to measure and draw angles on a compass rose is the *goniometer*. An excellent activity that incorporates the use of the goniometer as students explore the compass rose and the naming of airport runways can be found in the article "You Are Cleared to Land" (Billstein 1998).

Ask students what angle measure is used to describe the direction

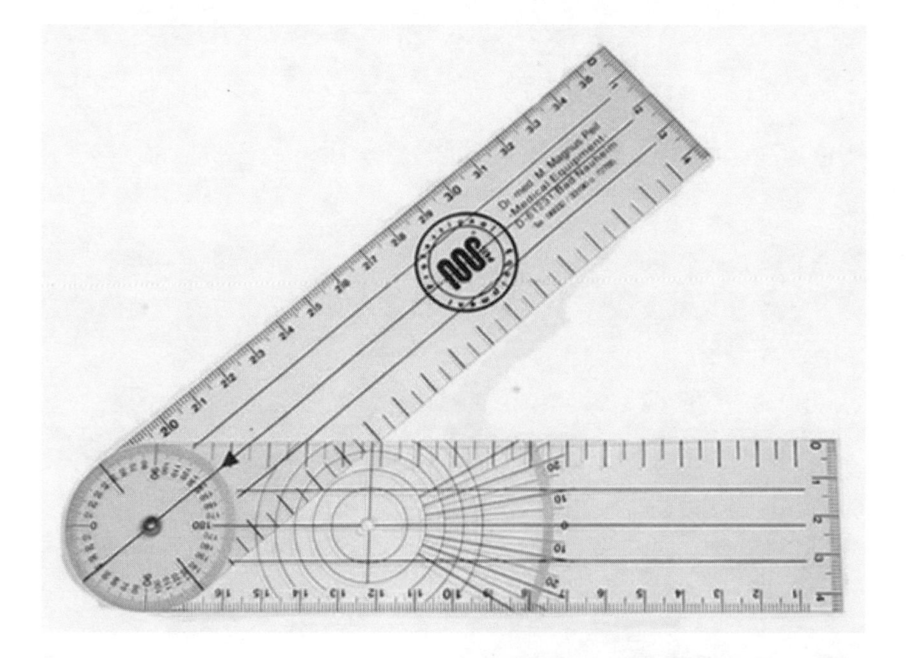

north. Ask them what other angle measure could be used to describe the same location.

Concluding the Activity

This activity offers an opportunity to check students' understanding of the process of measuring and drawing angles. Students can be asked several questions regarding directions on their compass roses to determine their levels of understanding. Students should also reflect on their understanding by completing a log sheet for this activity.

Tell students that they will use what they have learned while making the compass rose to explore how runways are named.

ACTIVITY 2

Exploring the Meaning of Runway Names

Important Mathematical Ideas

The procedure for naming airport runways supplies the context for further study of angle measures. This activity also presents opportunities for students to explore parallel and perpendicular lines and rounding.

Mission

Students use angle-measuring tools and rounding to determine runway names. Students try to identify a generalization to describe runways with opposite approaches.

Materials and Equipment

Compass roses, airport diagrams, and protractors

Conducting the Activity

Have students return to their mission teams. Give each team a compass rose with a runway drawn on it (see below). Tell the class that when an airplane comes in for a landing, its heading is determined by the compass rose. Ask students to work in their teams to determine the heading of the airplane coming in to land on the runway pictured on the compass rose.

Ask teams to share their answers, which will vary. Some students will say that the airplane's heading is between 70 degrees and 80 degrees, and some will say that it is between 250 degrees and 260 degrees. Students should soon see that both answers are correct, depending on which direction they assumed that the airplane was traveling, toward the west-southwest or toward the east-northeast.

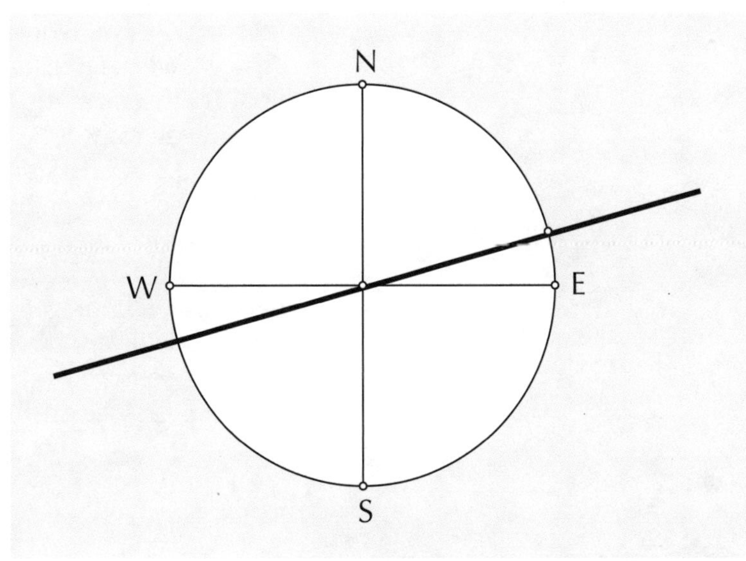

Developing the Activity

Explain that this dilemma and the range of angle measures are both taken into account when naming runways. Give each team the airport taxi diagram for Jacksonville International Airport (JAX) (see appendix, page A-22). The airport has two nonparallel runways, marked by thick, dark, long lines. Check to make sure that students can identify the runways on the airport diagram. Tell teams that the runways each have two names. One

NCTM Standards

Instructional programs from prekindergarten through grade 12 should enable all students to—

Algebra

Understand patterns, relations, and functions

Represent and analyze mathematical situations and structures using algebraic symbols

Measurement

Apply appropriate techniques, tools, and formulas to determine measurements

Internet Connection

Airport taxi diagrams can be found at the following Web site: www.aopa.org/asf/taxi/

runway is called 13 and 31, and the other runway is called 7 and 25. Have students use what they have learned about the compass rose, along with their protractors, to come up with conjectures regarding how the runways are named.

A runway is named using the heading an airplane is on as it approaches or takes off on that runway. The angle is rounded to the nearest ten, and the zero in the ones place is left off. Given time and some prompting, students should be able to come up with this method for naming runways.

Give students the airport taxi diagram for Syracuse Hancock International Airport without the runway names (see appendix, page A-23), and have them work in teams to determine the correct names. The runways are named 10-28 and 15-33. Students should discuss reasons for any differences in their results.

Concluding the Activity

Give each team the diagram for Las Vegas/McCarran International Airport without the runway names (see appendix, page A-20), and have them name the runways. Students should find that two pairs of runways have the same name. Lead a discussion about the problems that might be caused by having two runways with the same name. Ask students to come up with schemes for giving different names to parallel runways. Students' responses may be quite creative, but parallel runways are actually named using the same angle measures, followed by an R for the runway on the right and an L for the runway on the left. If an airport has a third runway, C is used to denote the center runway. When an airport has four parallel runways, as is the case in Atlanta, the runways are named to the next nearest ten-degree mark. For instance, in Atlanta, the runways are named 8L, 8R, 9L, and 9R.

Extending the Activity

Give students the diagram for Chicago-O'Hare International Airport (see appendix, page A-21), and ask how many different runway directions are shown. Students may count four or eight runways. If the answer given is four, ask students about landing in the opposite direction. For instance, the heading is different if an airplane lands the opposite way on runway 14R; the airplane would actually be landing on runway 32L. Have students use the diagrams to determine a rule for finding the name of the runway going in the direction opposite the one given. For instance, if the students know that a runway is named 18, how could they determine the name of the runway in the opposite direction without using the compass rose or looking at the diagram? Answers will vary but should involve adding or subtracting 180 degrees or 18. Encourage students to state their answers algebraically, using such expressions as $n + 18$ or $n^\circ + 180^\circ$.

ACTIVITY 3

Which Way Should I Land?

Important Mathematical Ideas

Students explore the ways in which wind speed and direction influence how airplanes land and whether the Shuttle is able to land on schedule.

Mission

Students use information about wind speed in knots to determine the conditions under which airplanes can land.

Materials and Equipment

Airport diagrams

Launching the Activity

Have students reexamine the Jacksonville International Airport diagram (see appendix, page A-22). Ask which runway a pilot would land on and which runway a pilot would take off from if the wind was blowing at 10 knots from the east. Define the term *knot* as speed in nautical miles. A nautical mile is slightly longer than a statute mile, which is the measure most commonly referred to on land. One nautical mile is approximately 1.15 statute miles; thus, if the wind is blowing at 10 knots, it is blowing at approximately 11.5 miles per hour.

Students will probably say that an airplane should land into the wind and take off with the wind, and they may be interested to learn that pilots try to both land and take off into the wind. Students may try to launch their paper airplanes with the wind and into the wind to compare the results.

Developing the Activity

If the wind is blowing strongly from a direction that is not parallel or almost parallel to any of the runways at an airport, airplanes are sometimes not able to land at that airport. Give students the diagram for Orlando International Airport (see appendix, page A-24). If the wind is blowing from the east or west at or above 30 knots, many airplanes are not allowed to land at this airport. Ask students to come up with explanations for this regulation. If an airplane is not allowed to land at Orlando, it may be diverted to Tampa. Ask students what runway directions they might expect to find at the Tampa airport.

Challenge students to determine the approximate speed in statue miles per hour of the wind if it is blowing at 30 knots.

Group students in mission teams to reexamine each of the airport diagrams used in "Exploring the Meaning of Runway Names" and to

NCTM Standards

Instructional programs from prekindergarten through grade 12 should enable all students to—

Number and Operations

Compute fluently and make reasonable estimates

Geometry

Specify locations and describe spatial relationships using coordinate geometry and other representational systems determine wind directions that would prevent airplanes from landing if winds are very strong. Students should share and compare their findings as a class.

Even the Shuttle landings are controlled by wind conditions. Students may be interested to learn that decisions regarding whether the Shuttle will be able to land are made just 70 to 90 minutes before landing time. If the crosswind is in excess of 15 knots during the day or 12 knots at night, if the headwind exceeds 25 knots, or if the tailwind exceeds 15 knots, the Shuttle is not able to land. Have students determine the approximate speed in statue miles per hour for these wind restrictions.

Concluding the Activity

Students should note that the headwind can be stronger than wind from any other direction and still allow for a Shuttle landing. Lead a discussion about how this information might be used to choose the direction for the runway. How might wind information be used in determining runway designs at commercial airports? This question is explored in activity 4.

Kennedy Space Center Shuttle landing facility

ACTIVITY 4

Making and Using a Windsock

This activity must be conducted over several days. On the first day, students construct the windsocks, design the experiment, and choose locations for data collection. Students then collect data for several days at different times during the day. Data collection does not necessarily need to be conducted during class. The activity is concluded with students' presentations of their findings.

Important Mathematical Ideas

Determining appropriate designs for experiments is imperative to collecting useful data. Students work in teams to create a tool to collect wind-direction data, then formulate data-collection plans to systematically collect wind-direction information and determine the prevailing wind direction at their school.

Mission

Students design and construct windsocks and use them to determine the prevailing wind direction at their school.

Materials and Equipment

Ripstop or other lightweight fabric, glue, scissors, wire, corks or blocks, swivels, and compasses

Launching the Activity

Show students the diagram for Orlando International Airport (see appendix, page A-24). Have students compare this diagram with the one for Chicago-O'Hare International Airport (see appendix, page A-21), and ask them to brainstorm ideas to explain why the runways are placed so differently. If students do not suggest wind as a factor on their own, you might remind them of what they learned in the preceding lesson about wind limitations on landing. Ask students what they might conclude regarding the wind in Orlando and in Chicago, noting that Chicago is sometimes referred to as the Windy City.

Ask students what they think the prevailing wind direction is at their school and how they might test their conjectures. Students may be familiar with windsocks if they have traveled in airplanes or visited an airport and seen windsocks on the runways. Show students a picture of a windsock if they are unfamiliar with this device.

Explain that each team will make a windsock and design an experiment to collect data on the prevailing wind direction at the school. Tell students that they will need to make a pattern to cut out the material

NCTM Standards

Instructional programs from prekindergarten through grade 12 should enable all students to—

Geometry

Analyze characteristics and properties of two- and three-dimensional geometric shapes and develop mathematical arguments about geometric relationships

Data Analysis and Probability

Formulate questions that can be addressed with data and collect, organize, and display relevant data to answer them

Select and use appropriate statistical methods to analyze data

Develop and evaluate inferences and predictions that are based on data

that will form the windsock. Give each team a few large pieces of paper to experiment with making and testing patterns for windsocks. Ask students what three-dimensional shape their windsock approximates. It is similar to a cone. This discussion may offer a lead-in to investigating two-dimensional patterns for other solids. Although students will likely find only one workable pattern for the windsock, several patterns can be used to form a cube, as well as pyramids and other prisms.

Developing the Activity

Once their patterns for the windsock are approved, the design teams should cut out material for the final product. Lightweight material should be used in areas that experience light winds, and heavier material may be used in areas that have heavy winds.

While some members of the design team cut out and glue the material, other members should bend the wire and attach it to the swivel, then attach the swivel to a straight piece of wire inserted into the cork or block.

Once the windsocks are constructed, the design teams need to decide how they will collect wind-direction data. Students should include their data-collection plans with justifications in their logbooks. In approving the data-collection plans, be sure to check that the teams will collect data at more than one time during the day. Wind intensity and direction often change from morning to afternoon. Students also need to use a compass to determine the direction of the wind in relation to magnetic north in the area.

Students must keep careful records during the data-collection period. The design teams will subsequently use their data to present a report to the class regarding the prevailing wind direction at their school. The reports should include descriptions of how the data were collected, as well as graphs of the data to support students' conclusions. This information is used in the following activity.

Concluding the Activity

Following the reports, lead a class discussion highlighting the similarities and differences in data-collection plans and team results regarding prevailing wind direction. Students should focus on how the plan for data collection could have affected the conclusions made by the teams.

Extending the Activity

This discussion could lead to a much broader examination of how information about public opinion on current topics is gathered and how conclusions are made on the basis of the data collected.

ACTIVITY 5

Designing Runways for Your School

Important Mathematical Ideas

In preceding activities, students used protractors and the compass rose to determine the directions of runways. In this activity, students use the same tools to draw and name runways in a given direction. The resulting airport drawings will probably vary greatly; this variance is acceptable as long as students can justify their choices.

Mission

Students apply what they have learned about airport runways and the prevailing wind direction at their school to create an airport for their locality.

Materials and Equipment

Compass rose, straightedge, protractor, poster board, markers, airport diagrams from local airports, and log sheets (see appendix, page A-2)

Launching the Activity

In this activity, students use the data collected on local prevailing wind direction and the knowledge they gained regarding how runways are named to design an airport for their locality. The mission teams will determine the complexity of their runway diagrams, but each team must justify the choices made in designing its airport. Further, all runways must be labeled properly.

Once the runways are designed, students might enjoy transferring their airport diagrams to poster board for presentation and subsequent display around the classroom. Comparing the different runway diagrams should make for an interesting discussion.

Concluding the Activity

Operators at local airports are often willing to give outdated runway diagrams to schools. If possible, bring these diagrams to class or locate the diagrams on the Internet to share with the class. Students will be interested in comparing their diagrams with those of an actual airport in the area. The similarities and differences in the student-generated designs and the designs of the airport may serve as motivating topics for discussion.

Students should record the similarities and differences in their teams' designs and the design of the local airport, along with possible reasons for the differences, in their logbooks. Students may also wish to adapt their runway designs to serve as mission patches or logbook covers.

NCTM Standards

Instructional programs from prekindergarten through grade 12 should enable all students to—

Measurement

Apply appropriate techniques, tools, and formulas to determine measurements

Extending the Activity

After students have gained a basic understanding of runway design, they may be eager to talk to a pilot about the knowledge they have acquired and to learn the answers to questions they may have. Invite a commercial or private pilot to the class to talk about flight plans, different airports, and his or her experiences in flying planes.

Have each team write two questions to ask the pilot and turn the questions in to you for approval before the visit. You may want to compile the questions for the pilot to read in preparation for his or her visit.

Begin the visit with a presentation by the pilot, then have a spokesperson for each team take turns asking the students' questions. Be sure to allow time for additional questions at the end of the visit. Students should record what they learned from this visit, along with questions they still have, on their log sheets.

Pompano Air Park, Florida

International Space Station and Other Earth-Observing Satellites

Each November 2nd, the International Space Station (ISS) sets new records for continuous international human presence in orbit. Assembled by astronauts, the ISS is becoming the largest, most sophisticated spacecraft ever constructed. When it is complete, the station will represent a project of unprecedented size and complexity in space. Led by the United States, the ISS program draws on the scientific

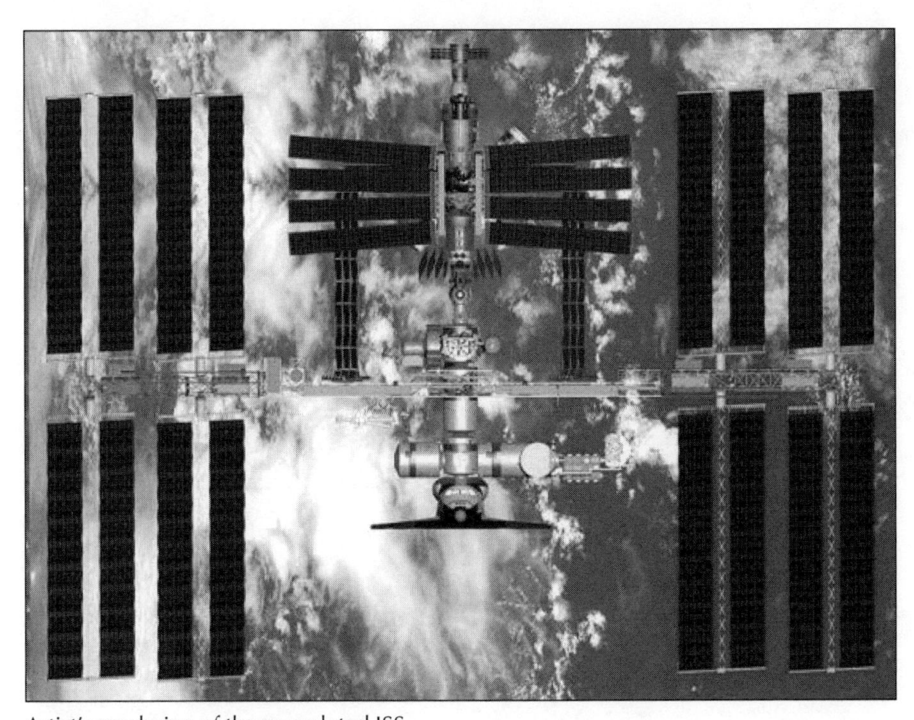
Artist's rendering of the completed ISS

and technological resources of fifteen other nations: Canada, Japan, Russia, Brazil, and eleven nations of the European Space Agency.

The United States has the responsibility for developing and ultimately operating major elements and systems aboard the station. The U.S. elements include three connecting modules, or nodes; a laboratory module; truss segments; four solar arrays; a habitation module; three mating adapters; a cupola; an unpressurized logistics carrier; and a centrifuge module. The various systems being developed by the United States include thermal control; life support; guidance, navigation, and control; data handling; power; and communications and tracking, along with ground-operations facilities and launch-site-processing facilities.

The international partners, Canada, Japan, the European Space Agency, and Russia, will contribute essential elements to the ISS, as well.

Using Earth-Observing System (EOS) satellites, NASA is playing a central role in the effort to understand the complexities of our planet. Earth is a dynamic planet. Change is constantly occurring, and the consequences of the changes are important to the future of the human race. NASA is contributing to humankind's understanding of Earth through the program known as Mission to Planet Earth. Since 1991, NASA has supported scientific research that focuses on specific aspects of Earth's environment. These efforts are often conducted in cooperation with other federal agencies and scientists from other nations.

The EOS program involves flights of a diverse set of instruments on a number of spacecraft over a fifteen-year period. Using advanced computing systems, scientists will compile data about Earth from all available sources and attempt to model Earth as a global system. Particular attention is being given to how human activities affect the planet (NASA 1995, p. 25).

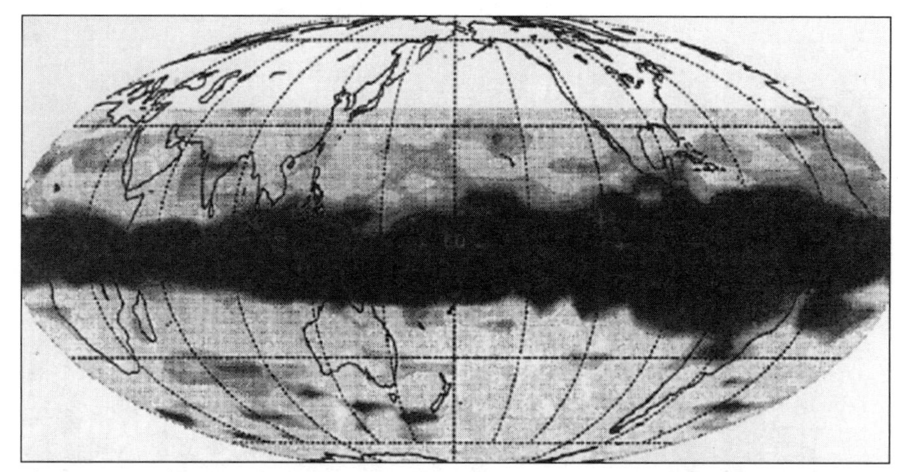

The eruption of Mount Pinatubo in the Philippines released sulfur dioxide, which appears in this 1991 diagram as a dark belt concentrated around the tropics

International Space Station

Throughout the history of the space program, considerable activity in space exploration has taken place around Earth. The early suborbital flights of Project Mercury began in 1958, and six manned flights occurred from 1961 to 1963. Alan Shepherd made the first U.S. suborbital flight, which lasted just fifteen minutes. Today, the ISS orbits Earth, and astronaut scientists can remain in orbit for weeks to perform experiments and study our planet.

The ISS is a platform in space where scientists and engineers from many countries can perform complex, long-term, replicable experiments. This space station is 131 feet high by 290 feet wide by 356 feet long; it is the largest man-made object in space and will eventually have a mass of about 1,040,000 pounds. It is about the same size as two Boeing 747 airplanes. Because of the immensity and complexity of the ISS, 45 launches and more than 1,700 hours of space walks are required to assemble it. Fully constructed, the ISS will be visible to more than 90 percent of the world's population.

The station will be in an orbit with an altitude of 250 statute miles and an inclination of 51.6 degrees. This orbit allows the station to be reached by the launch vehicles of all the international partners to facilitate the delivery of crews and supplies. The orbit also offers excellent Earth observations, with coverage of 85 percent of the globe and overflight of 95 percent of the population.

The *Mission Mathematics* activities associated with the ISS are designed to give students a better idea of how materials are delivered to the station and promote thinking about some of the considerations for living aboard the ISS. The activities are as follows:

- "Cargo Bay on the Space Shuttle": Students find common cylindrical containers that have similar proportions to the Shuttle cargo bay.
- "Twenty-Foot Equivalent Units": Students learn about a unit of measure used in shipping and compare the size of different means of shipping, including the Shuttle, using this measure.
- "Leonardo, Raffaello, and Donatello": The focus of this activity is on the best shape for "filling" the Shuttle cargo bay.
- "Balancing 'Linear Objects'": Students learn firsthand about the concept of *center of mass* (CM).
- "Does the CM Move?": Students think about adding mass to the ISS by experimenting with the CM .
- "Finding the CM of 'Flat Objects'": Students experiment more with the concept of CM.

Internet Connection

Crew training for missions to the ISS involves many hours in the neutral buoyancy tank at Johnson Space Center. For a lesson plan related to neutral buoyancy, access the following Web site:

www.quest.arc.nasa.gov/neuron/teachers/stellar/Neutral.html

For background information on the Neutral Buoyancy Simulator at Marshall Space Center, consult this Web site:

www.msfc.nasa.gov/NEWSROOM/background/facts/nbs.htm

More information about the Sonny Carter Training Facility Neutral Buoyancy Laboratory at Johnson Space Center is available at the following site:

www.jsc.nasa.gov/

ACTIVITY 1

Cargo Bay on the Space Shuttle

NCTM Standards

Instructional programs from prekindergarten through grade 12 should enable all students to—

Measurement

Apply appropriate techniques, tools, and formulas to determine measurements

Geometry

Use visualization, spatial reasoning, and geometric modeling to solve problems

Explorers have always been great planners. Meriwether Lewis planned painstakingly for the expedition across the West with William Clark. He had to estimate what provisions he would need on the long journey and how much of each item he could carry. He even invented a boat with a metal frame and a "skin" of canvas to carry the provisions for river travel. Today, a great deal of scientific work is done in places where supplies are not readily available. For example, scientists need to plan for provisions for research outposts in the Antarctic. The harsh climate of Antarctica does not allow for easy delivery of supplies.

As it is for Earth-based exploration, constant resupplying of fuel, food, and other cargo is important for the ISS. Because the environment of space, even in the ISS, is not conducive to long-term habitation, the astronauts are also cycled on and off the ISS regularly. This effort requires intense planning to ensure that the astronauts are safe and that the ISS is fully operational. NASA and other space professionals on Earth do this planning. This activity explores some of the problems planners might encounter as they look for ways to keep the ISS well supplied.

Important Mathematical Ideas

Students use their knowledge of three-dimensional shapes and their ability to measure the volume of containers.

Mission

Students try to visualize the shape of the cargo bay of the Shuttle orbiter.

Materials and Equipment

A collection of common cylindrical objects, such as tennis-ball cans, cooked-oats boxes, baking-powder cans, potato-chip containers, and so on; rulers; protractors; and compasses

Launching the Activity

Visualizing the shape of a cylindrical object is sometimes difficult even for adults. Ask students to bring in empty cylindrical containers to make a classroom collection. Before discussing the Space Shuttle, have students make a table of the diameters and heights of the cylindrical objects in the collection, as shown on the facing page.

Developing the Activity

Before exploring the different volumes of the cylinders in the table, tell students that the Space Shuttle cargo bay is 60 feet long and has a diameter of 15 feet. Ask students to talk with partners about which one

Object	Height	Diameter	Volume
Tennis-ball can	8.5 in	3 in	
Quaker Oats box	7 in	4 in	
Morton Salt box	5.375 in	3.25 in	

of the cylindrical objects in the class collection has a shape most like that of the Space Shuttle cargo bay. Have students write their choices on sticky notes and "post" their votes on the chalkboard in the form of a bar graph. Discuss the voting, and ask students to justify their answers.

Concluding the Activity

As justification, some students may mention the ratios of the diameters of the cylinders to their heights. For example, one pair of students may respond, "The ratio of the diameter of the cargo bay to the height of the bay is 15 to 60. The measurements of the tennis-ball can [or other classroom cylinder] are similar to those measurements in inches." With some good fortune, a pair of students may even bring up ratios and proportions, giving the teacher the opportunity to introduce the concepts of ratio and equivalent ratios.

ACTIVITY 2

Twenty-Foot Equivalent Units

NCTM Standards

Instructional programs from prekindergarten through grade 12 should enable all students to—

Measurement

Apply appropriate techniques, tools, and formulas to determine measurements

Geometry

Use visualization, spatial reasoning, and geometric modeling to solve problems

In the shipping industry, standard containers are used to ship many products. *Twenty-foot equivalent unit* (TEU) is the name given to these standardized containers because they are designed in the shape of a rectangular prism measuring 20 feet long, 8 feet high, and 8 feet deep. These containers fit on flatbed truck trailers, on train flat cars, and in large container ships. Some of the biggest container ships can carry up to 8000 TEUs. These ships are almost as long as four football fields and nearly as wide as one football field. How do these large ships compare with the Space Shuttle?

Important Mathematical Ideas

Students use their knowledge of three-dimensional shapes and their ability to measure the volume of containers.

Mission

Students explore the capacity of the Shuttle orbiter using TEUs.

Materials and Equipment

A collection of common cylindrical objects, such as tennis-ball cans, cereal boxes, baking-powder cans, potato-chip containers, and so on; rulers; protractors; and compasses

Launching the Activity

Initiate a discussion of students' experiences seeing TEUs in their community. They may have seen a freight train carrying these storage units on flatbed cars. They may have seen eighteen-wheeler trucks carrying these units, or they may have seen them used at construction sites as storage units.

Ask students, "How many TEUs could fit into the cargo bay of the Space Shuttle?" but caution them that this question requires them to "put a square peg in a round hole." Explore this idea by asking, "What shape is the cargo bay? What shape is a TEU? If we stood a cylinder and a rectangular prism on end and used them like rubber stamps, what shapes would be produced?"

Teaching Note

Some students may notice that the diagonal of the best square is the same as the diameter of the circle.

Developing the Activity

Divide students into mission teams, and have them select the container from the first activity that they found to be the best model of the cargo bay. Remind students to think about containers shaped like TEUs and to try to estimate the size of the cargo bay. Then have each student on the team perform the following steps:

- Stand the model on end, and trace around its circular base.
- Try to draw a square that is sized so that its four corners touch the circle; cut out these squares.
- Measure the length of the sides of the cutout squares.

As a team, have students answer the following questions:

- If one of your squares was the end of a TEU-like container that was 20 inches long, what would be the volume of this container with rectangular ends?
- Is the volume of these containers smaller or larger than the volume of the best model of the cargo bay?
- Did all four corners of your square touch the circle? If not, tell students to try again. They may need to make several tries to find just the right square. Students should be ready to share how they finally made the best fit.
- If the circle was the end of a 20-inch container, what would be the volume of that shape?
- How much larger is the best model of the cargo bay with the cylindrical ends than the "TEU" with rectangular-shaped ends?

Teaching Note

The formula for the volume of a cylinder is $\pi r^2 h$.

Concluding the Activity

Students should discover that if the Earth-based standard shipping container, that is, the TEU, was used in the Space Shuttle cargo bay, a great deal of space would be wasted. Have teams try to identify a better storage-unit shape for the Space Shuttle cargo bay.

ACTIVITY 3

Leonardo, Raffaello, and Donatello

NCTM Standards

Instructional programs from prekindergarten through grade 12 should enable all students to—

Measurement

Understand measurable attributes of objects and the units, systems, and processes of measurement

Algebra

Understand patterns, relations, and functions

Analyze change in various contexts

Use mathematical models to represent and understand quantitative relationships

Geometry

Use visualization, spatial reasoning, and geometric modeling to solve problems

NASA Administrator Daniel S. Goldin and the president of the Italian Space Agency (ASI), Sergio De Julio, met at NASA's Kennedy Space Center in Florida on Friday, September 25, 1998, for a ceremonial event transferring the "Leonardo" Multipurpose Logistics Module (MPLM) from the ASI to NASA.

The MPLM, a reusable logistics carrier, is the primary delivery system used to resupply the ISS and return station cargo requiring a pressurized environment. It is one of Italy's major contributions to the ISS program. The cylindrical module is approximately 21 feet long and 15 feet in diameter and weighs almost 4.5 tons.

Leonardo is the first MPLM. Raffaello and Donatello are MPLMs, too. MPLMs are carried to the ISS aboard the Space Shuttle and are temporarily docked to the station. Once in orbit, the MPLMs provide a working environment for a crew of two. Each MPLM can carry up to 20,000 pounds of supplies, materials, and equipment for science experiments, spare parts, and other logistical components for the ISS.

Important Mathematical Ideas

Students intuitively explore the notion of limit as they try different shapes of "containers" for the cargo bay of the Shuttle.

Mission

Students continue to explore the shapes of containers for the cargo bay of the Shuttle.

Materials and Equipment

For each student, a compass, a protractor, and a log sheet (see appendix, page A-2)

Launching the Activity

Explain to students that to fill more of the Space Shuttle cargo bay, the ASI built "containers" that fit better than rectangular ones. How did the designers decide what shape would fit the Shuttle best? Remind students to think again about the fit of a square inside a circle.

Begin the activity by defining *central angles:* "Angles formed by intersecting diameters of a circle are called *central angles* of the circle."

Remind students that a circle measures 360 degrees. Have them talk with their partners or team members to answer the question "What is the measure of the central angles formed by the diagonals of the largest possible square drawn inside a circle?" Students should be prepared to justify their answers.

Developing the Activity

The measurement 360° is divisible by 4, making each of the angles 90°; 360° is also divisible by 6. Ask students, "What shape do we get if we draw six equal central angles and connect the endpoints to form a polygon? If the Space Shuttle cargo container were a hexagon, would more or less space be wasted?"

Continue the discussion by asking the following questions:

- What other numbers can divide 360 equally?
- If we make equal central angles using various divisors of 360, what are the measures of the central angles and what shapes are produced using these angles?
- Which shape wastes the least amount of space?
- What happens to the wasted space as the number of sides on the container increases?

Concluding the Activity

The number of sides on the Leonardo, Raffaello, and Donatello MPLMs is fourteen, which is not an integral divisor of 360. If your class is studying fractions or decimals when doing this activity, students can use many more numbers to create shipping containers for the Space Shuttle cargo bay.

Students should recognize the progression of containers from squares to many-sided shapes. As the number of sides of the polygons increases, the shapes approach the shape of a circle. In three dimensions, the shapes of the polyhedra become progressively closer approximations of the shape of the cylinder, that is, the cargo bay. This activity offers a good intuitive introduction to the concept of limit. By looking at the wasted space, students can see that a limit of zero wasted space is approached as the number of sides increases.

Students should reflect on the mathematical concepts explored in this activity as they complete the log sheet for their logbooks.

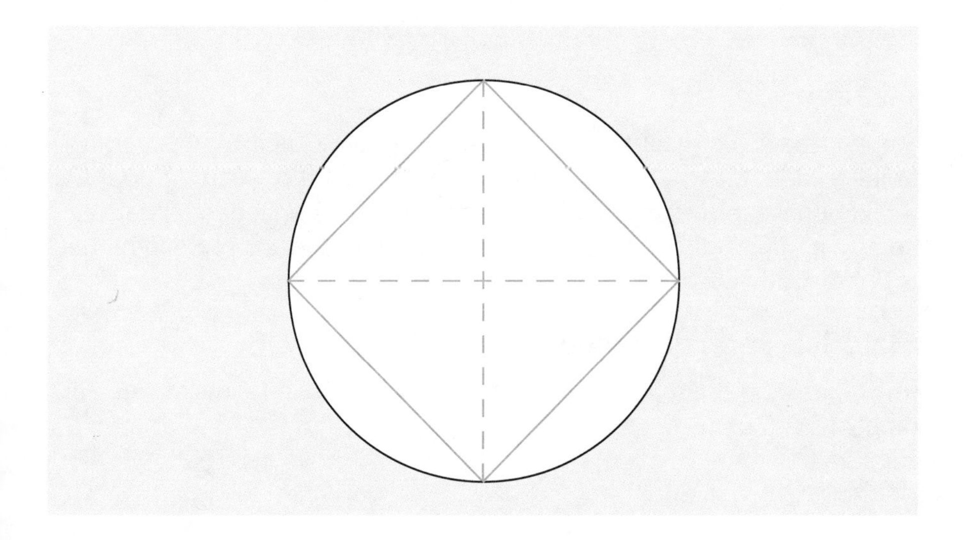

ACTIVITY 4

Balancing "Linear" Objects

NCTM Standards

Instructional programs from prekindergarten through grade 12 should enable all students to—

Measurement

Understand measurable attributes of objects and the units, systems, and processes of measurement

When early explorers from Europe sailed to the New World, they used lead and water as ballast. These explorers knew that loading something heavy, *ballast,* in the bottom of their ships made the ships more stable. That is, the ships were less likely to capsize in high winds and seas. The explorers hoped that the ballast on the return trip to Europe would be gold and silver from the New World.

Stability is important to NASA scientists, too. Stable satellites are easier to keep in prescribed orbits, and images of a specific part of Earth are more reliable when the satellite is stable. A tumbling satellite does not produce accurate images for use by scientists.

The center of mass (CM) is an important concept in science. Although the mathematics related to the CM of irregularly shaped objects is beyond most middle school students, an exploration of this concept using grade-level mathematics can be valuable.

Important Mathematical Ideas

Theoretically, the concept of CM of a "linear" object involves ratio; however, experimentation with classroom materials often allows only approximations of these ratios. Students use many middle school mathematics skills in this activity and make inferences about the location of the CM of an object.

The CM is defined as that point in a given object that represents the mean position of the matter in the body of the object.

All matter, regardless of its size, mass, or shape, has a point inside called the *center of mass* (CM). The CM is the exact spot where all the mass of that object is perfectly balanced.

Mission

Students have a concept of middle. In this activity, their understanding of the middle is expanded to begin to encompass the meaning of CM.

Students explore relationships involved in the concept of CM using common objects that have one dimension, in this instance, length, that is significantly greater than the other two dimensions.

Materials and Equipment

Rulers, unsharpened pencils with erasers, a broom, a tennis racket, and other "linear" objects

Launching the Activity

Students should begin to build the concept of CM in this activity, which makes use of common objects. The term linear is placed in quotes in the title of the activity to indicate its use as a descriptor of common objects, not as a mathematical term.

Show students how to balance a ruler on the tips of their index fingers by putting one finger at each end. Next, have them slide their fingers slowly toward the middle of the ruler. Make sure each finger moves at the same speed. When their fingers come together and the ruler balances, they have reached the CM for the ruler.

Next have students try to balance the ruler at the CM using the eraser end of a pencil. Note that the CM is on or near the middle of the ruler.

Not all linear objects have a CM in the middle as determined by a ruler or meterstick. Have students try balancing a new pencil or a broom in the same manner as they did the ruler. The CM is not in the measured middle of the object.

Concluding the Activity

As a class, make a list of linear objects that have the CM in the measured middle of the object and a list of linear objects that have the CM closer to one end of the object.

Object	CM in the measured middle	CM is nearer to one end of the object
Ruler	✓	
Broom		✓

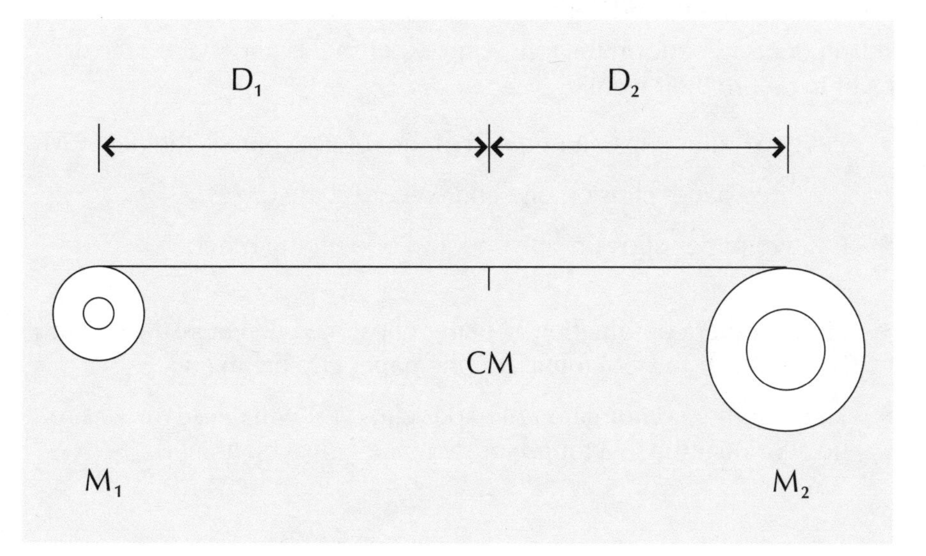

ACTIVITY 5

Does the CM Move?

NCTM Standards

Instructional programs from prekindergarten through grade 12 should enable all students to—

Measurement

Understand measurable attributes of objects and the units, systems, and processes of measurement

Algebra

Understand patterns, relations, and functions

Analyze change in various contexts

Use mathematical models to represent and understand quantitative relationships

When new mass is added to one end of an object, the CM is not in the same position. This problem must be considered when loading ships or the ISS. As the Space Shuttle arrives at the ISS, it brings tons of supplies and is prepared to take loads of trash back to Earth. Balancing these loads is important to the stability of the ISS and the Shuttle.

Important Mathematical Ideas

After understanding that different objects have CMs in different locations, students are prepared to experiment with the location of the CM of an object.

Mission

Students explore relationships involved in the concept of CM of one object as they "alter" the object.

Materials and Equipment

Index cards, paper clips, scissors, ruler

Launching the Activity

Cut a 5-by-8-inch index card into five 1-inch strips, each 8 inches long. Share the strips with other students. Use a standard ruler to mark off inches, half inches, and quarter inches on one of the paper strips. The result is a usable 8-inch ruler. Ask students to determine the CM for this new paper ruler. Next have them clip one paper clip to one end of the ruler and predict the location of the CM when the paper clip is attached.

Concluding the Activity

Tell students to perform the following experiments and make a list or chart to record their results.

- Put two paper clips on one end of the ruler. Where is the new CM?
- Put three paper clips on one end. Where is the CM?
- Put two paper clips on one end and one on the other. Where is the CM?
- Find another combination of paper clips that will cause the CM to be at or near the CM found for one paper clip on an end.
- Find another combination of paper clips that will cause the CM to be at or near the CM found for two paper clips on an end.

ACTIVITY 6

Finding the CM of "Flat" Objects

So far, the exploration of CM has focused on linear objects, but most applications of CM must consider more than one dimension. Balancing the ISS requires thinking in more than one dimension.

Important Mathematical Ideas

Students have made inferences about the CM of linear objects. Next they are prepared to experiment with the location of the CM of a flat object. This activity requires the application of students' knowledge of graphing.

Mission

Students make inferences about the location of the CM of a flat object. The concept of the Cartesian plane is used to help students describe their ideas and questions.

Materials and Equipment

Index cards, graph paper, paper clips, scissors, glue, rulers, and log sheets (see appendix, page A-2)

Launching the Activity

Have each student glue graph paper to an index card so that the grid lines align with the edges of the cards. Students should then cut off the excess graph paper and label each corner of the card with roman numerals, starting at the upper right and moving counterclockwise. Then have them turn the index cards so that the graph paper is facing down.

Ask students to try to find the CM of the index card by balancing the card on the eraser of a pencil. Note the CM and mark it with an *X*.

Next have students perform the following steps: On the graph-paper side, determine and mark the middle of each edge of the index card. Create an *x*-axis and a *y*-axis by connecting the middle points on opposite sides of the index card. The CM should be near the origin. The corner labeled I is the first quadrant of the coordinate system. Fill in some of the numbers on the axes of this coordinate system.

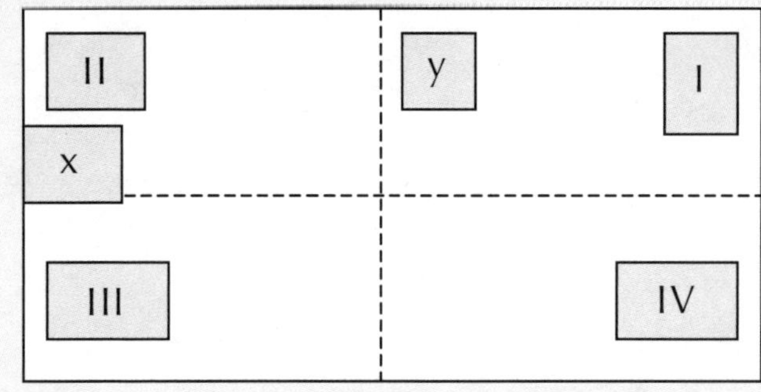

Developing the Activity

To explore what happens when paper clips are added to the index card, place a paper clip on one corner of the index card, find the CM, and mark it with an *X*. Move the paper clip to the

NCTM Standards

Instructional programs from prekindergarten through grade 12 should enable all students to—

Measurement

Understand measurable attributes of objects and the units, systems, and processes of measurement

Algebra

Understand patterns, relations, and functions

Analyze change in various contexts

Use mathematical models to represent and understand quantitative relationships middle of one of the edges, and mark the CM. Try moving the paper clip to at least six other spots around the edge of the index card. Mark the CM for each move. Record the coordinates of each CM. Describe any patterns you see.

Predict what will happen when you move two paper clips around the edge of the index card. Add the two paper clips, and move them around. Record the coordinates of each CM. Describe any patterns you see.

Concluding the Activity

Using the prescribed materials in this activity is important to ensure that the experimental results will be as close as possible to the theoretical answers to questions about CM. A paper ruler with paper clips of equal weight at each end obviously has its CM in the middle. When two paper clips are used on one end, doubling the weight, the CM divides the distance between the paper clips by a ratio 1:2 in a way that makes the CM closer to the heavier mass. This principle also works for other ratios, but using wooden rulers or metersticks causes a distortion in the ratios because of the mass of the stick itself. For this reason, using light-weight material to experiment to find an unknown CM is desirable.

To conclude this activity, have students formulate their own questions about the concept of CM and record these questions on their log sheets. Students will often ask "What if ...?" questions. In the spirit of inquiry, encourage students to explore further and attempt to answer their own questions. Students may also wish to represent their new knowledge of CM on mission patches or logbook covers.

Extending the Activity

Some of the students' questions may focus on three-dimensional objects, such as "Does my skateboard have a CM?" or "Where is the CM of my bicycle?"

Finding the CM of three-dimensional objects is the next step in helping middle school students extend their understanding of the concept. Shoe boxes, plastic storage boxes, and oatmeal containers are suggested objects for initial exploration of CM in three dimensions. Using the intersection of lines, have students connect balance points on each "side" of the objects.

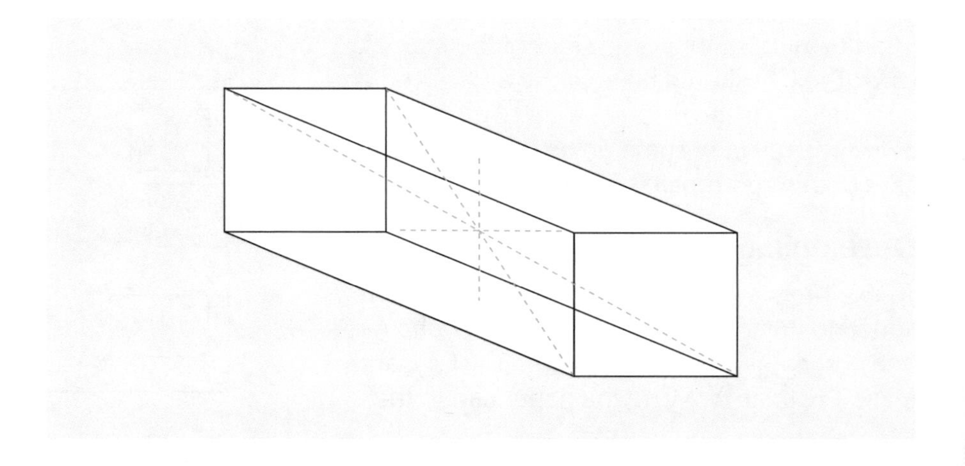

Other Earth-Orbiting Satellites

Weather satellites are important for collecting data about Earth. These unmanned spacecraft carry a variety of sensory equipment that scans Earth and electronically communicates the data back to scientists. Two types of environmental satellite systems named for their orbital characteristics, geo-stationary satellites and polar-orbiting satellites, provide the data for scientists to study our environment.

Geo-stationary satellites orbit at a speed that enables them to remain over the same area of Earth. This position offers constant "viewing" of a specific area of Earth. The Geo-stationary Operational Environmental Satellites are an important system for the United States. These satellites are maneuvered into orbits that ensure constant coverage of the Western Hemisphere. They continually produce infrared images, weather charts, ice charts, and other data important to the study of Earth's environment.

Polar-orbiting satellites operate in orbits that cause them to pass almost over the poles of Earth. Many of these satellites operate in Sun-synchronous orbits. These Sun-synchronous orbits are designed so that the satellite passes over the same terrain at the same time each day. This orbital feature is important for the collection of data about Earth.

This module includes the following six activities:

- "Orbit Primer": Students determine the velocity of satellites in orbit.
- "How High Is It?": Using the circumference formula, students determine the height of a satellite in orbit.
- "Ice Masses": Students find the area of irregular figures represented as ice masses.
- "Migration of Whales": Students use scaled maps to determine distances.
- "Finding Whales": The Pythagorean theorem is used to calculate distances traveled by whales.
- "GPS": Students use latitude, longitude, and bearing in this activity to track someone moving between two points.

A satellite image of Hurricane Fran off the east coast of Florida

ACTIVITY 1

Orbit Primer

Engage the class in a large-group discussion about the satellite images they see on local television news broadcasts. In the discussion, help students distinguish among animations of Shuttle missions; photographs from the Space Shuttle; pictures from remote robotic explorations, such as *Pathfinder* on Mars; Hubble Space Telescope images; and satellite photographs of Earth. If possible, obtain videotape of a television weather report of a fast-moving storm that has been tracked and shown in nearly real time. If the videotape cannot be obtained, ask students to recall watching such events on television. As part of the discussion, ask students to speculate on how the television station obtained these images. Some students may know that some television weather graphics are made with satellite data.

This discussion can lead to an opportunity to introduce students to geo-stationary and polar-orbiting satellites. The idea of polar-orbiting satellites—those with orbits that take them above both poles—may be easier for students to understand. They seem to comprehend the notion of the ground track of polar-orbiting satellites. The reason may be that the movement of these satellites is similar to that of an airplane in the sky. Thus, discussing polar-orbiting satellites first may be desirable. The class discussion will determine whether this approach seems appropriate.

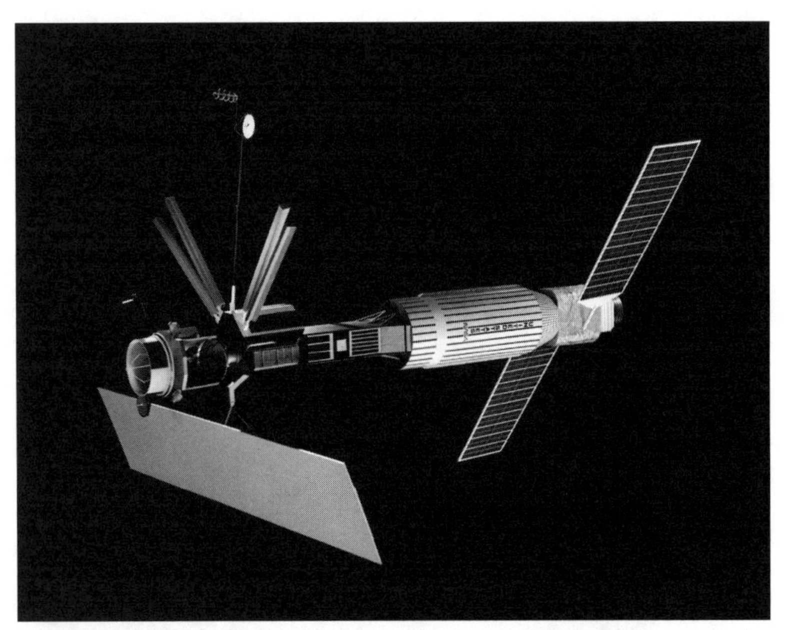

SEASAT was the first Earth-orbiting satellite designed for remote sensing of Earth's oceans.

Important Mathematical Ideas

Students use algebraic thinking to determine facts about orbiting satellites presented in a science context.

Materials and Equipment

Log sheets (see appendix, page A-2), calculators (optional)

Mission

In this activity, students explore the concept of orbits. The connection of Earth-orbiting satellites with the study of the environment is used to capture students' interest. Both geometric and algebraic concepts are presented to students in this science context.

Launching the Activity

Ask students the following question:

How do you determine the velocity of a polar-orbiting satellite when the altitude and the period of the orbit are known? For example, if a 750-mile-high satellite orbits Earth every 3 hours, 45 minutes, how fast is it traveling? In other words, what is the velocity of the satellite?

Allow students to work in small groups to draw pictures of the situation and attempt to think through the solution to the problem without doing the computation. Encourage them to write statements, such as "I would first change the 3 hours, 45 minutes to hours in decimal form" or "I would compute the distance that the satellite travels in one orbit of Earth by using the formula for the circumference of a circle."

Developing the Activity

Once the group has a plan to compute the velocity of the satellite, have each student in the group do the calculations to determine the velocity.

A sample solution path for the example is as follows:

Step 1. Change 3 hours, 45 minutes to

$$3\,\frac{45}{60}\text{ hours} = 3\,\frac{9}{12}\text{ hours}$$
$$= 3\,\frac{3}{4}\text{ hours}$$
$$= 3.75\text{ hours.}$$

Note that a fraction calculator can be used in this instance.

The radius of an orbit is calculated from the center of Earth. Consequently, the radius of an orbit equals the radius of Earth plus the altitude of the satellite above Earth.

Step 2. According to the formula $C = 2 \times \pi \times r$, the distance traveled is

$$2 \times \pi \times (3{,}960.5 + 750)\text{ mi} \approx 29{,}581\text{ mi.}$$

Step 3. This distance is traveled every 3.75 hours. Therefore, the speed, or velocity, of the satellite is

$$\frac{29{,}581\text{ mi}}{3.75\text{ h}} \approx 7{,}888\text{ mi/h.}$$

Have students calculate the velocity of satellites traveling at an altitude of 170 miles above Earth for a period of 1.5 hours (~17,293 mi/h) and 20,150 miles above Earth for a period of 20 hours (~7,570 mi/h).

Concluding the Activity

When students are sharing solutions, emphasize the path to the solution. Talk about the different approaches the students used to reach the same solution. Later, in a more formal algebra class, a teacher may focus on the equivalent forms of equations used to solve the problem. At this point, students should reflect on what they learned and record on their log sheets any questions that arose.

Teaching Note

To simplify the notion of orbits for middle school students, the orbits discussed in these activities are assumed to be circular.

Definition

The orbital period is the time required to complete one orbit around Earth.

ACTIVITY 2

How High Is the Satellite?

NCTM Standards

Instructional programs from prekindergarten through grade 12 should enable all students to—

Algebra

Represent and analyze mathematical situations and structures using algebraic symbols

Analyze change in various contexts

Geometry

Use visualization, spatial reasoning, and geometric modeling to solve problems

Measurement

Apply appropriate techniques, tools, and formulas to determine measurements

Important Mathematical Ideas

Students use algebraic thinking and measurement formulas to calculate the height of an orbiting satellite.

Materials and Equipment

Log sheets (see appendix, page A-2), calculators (optional)

Mission

Students extend their understanding of orbits in this multistep problem-solving activity. They use previous knowledge of the Earth's diameter and the formula for the circumference of a circle to approximate the radius (height) of an orbiting satellite.

Launching the Activity

Ask students, "How do you find the altitude of a geo-synchronous satellite that is orbiting Earth above the equator when the velocity of the satellite is known?" Using a value for the velocity, you might ask, "If a satellite is traveling at 6,900 miles per hour, how high must it be orbiting?"

This question is sometimes difficult for students because the problem has only one number. Putting students into groups to draw pictures and think about what is known about the problem will usually help them identify all the pertinent data. Drawing pictures will also facilitate their discussions.

Developing the Activity

One possible solution path is as follows:

Step 1. A geo-stationary satellite "orbits" Earth once every 24 hours. The period of the orbit makes it appear not to move. It is moving in unison with Earth's rotation.

Step 2. If a satellite travels 6,900 miles in 1 hour, how far does it travel in the 1-day orbit (a period of 1 day)?

$$6{,}900 \text{ mi/h} \times 24 \text{ h} = 165{,}600 \text{ mi}$$

Step 3. This result, 165,600 miles, is the circumference of a circle, and $C = 2 \times \pi \times r$. Thus,

$$2 \times \pi \times r = 165{,}600,$$

or

$$\pi \times r = \frac{165{,}600}{2}$$

$$= 82{,}800 \text{ mi},$$

and

$$r = \frac{82,800}{\pi}$$

$$\approx 26,369 \text{ miles.}$$

Step 4. This result, 26,369 miles, is the radius of a circle whose center is the center of Earth.

To determine the satellite's altitude above the surface of Earth, the radius of Earth (3,960.5 miles) must be subtracted from the radius of the circle. The altitude becomes 26,369 – 3,960.5, or 22,408.5, miles above Earth.

Concluding the Activity

Once again, an important aspect of the activity is having students share their solutions to this problem. Different students may see alternative paths to the same solution. When students share their thinking, some will gain new or heightened understanding of the topics under discussion.

For further discussion and practice, the problem can be changed to focus on determining the speed of the satellite when the height is known.

Have students record on their mission log sheets their new understanding of both the science and the mathematics of orbiting satellites.

ACTIVITY 3

Ice Masses

NCTM Standards

Instructional programs from prekindergarten through grade 12 should enable all students to—

Algebra

Represent and analyze mathematical situations and structures using algebraic symbols

Analyze change in various contexts

Geometry

Use visualization, spatial reasoning, and geometric modeling to solve problems

Measurement

Apply appropriate techniques, tools, and formulas to determine measurements

Sea levels have been rising slowly for nearly a century. Part of this change can be attributed to the melting of some glaciers and ice caps around the world. What we do not know is whether the world's two major ice sheets—in Greenland and Antarctica—are growing or melting. This knowledge is important because these areas are so large that their behavior can have global effects. As a point of reference, these two ice sheets contain approximately 75 percent of Earth's fresh water in frozen form. If they melted entirely, scientists estimate that sea levels would rise approximately 75 meters. This amount of water would flood much of the United States. Coastal states, including Florida, would be nearly covered by such a phenomenon. Although dramatic changes in the amount of ice in the glaciers, ice caps, and ice sheets of the world are not likely to occur in a short time, scientists must monitor these frozen bodies. Subtle changes in any of them may signal a long-term trend toward global warming or cooling.

NASA has been collecting data about ice masses for some time now. Early satellites in the 1960s and 1970s obtained valuable information, but the sophistication and power of the instruments on those early missions do not match those of the instruments available today. The main initiative of NASA's Mission to Planet Earth, the Earth-Observing System (EOS), will have instruments that are capable of collecting in-depth ice data even when clouds obscure the ice sheets from view. The series of EOS satellites will collect data over a period of fifteen to twenty years.

Glacier Bay, Alaska

This information will allow scientists to determine trends in data that indicate global warming or cooling.

Important Mathematical Ideas

By looking at classroom globes, maps in books, and NASA images, students realize the extent of ice in the world. They also see that the areas covered by ice are irregularly shaped. Students think about the attributes of polar ice that are important for scientists to monitor. The focus of the activity for all students is determining the area of an ice mass. Multiple approaches to finding the area of an irregular figure are considered.

Mission

Students suggest some attributes of the polar ice cap that would be important to measure.

Students devise and use various approximation techniques to estimate the area of a figure that represents a polar ice cap.

Materials and Equipment

A resource page for each student, graph paper, rulers, maps or photographs of the polar ice caps, and a globe

Launching the Activity

Introduce the activity by discussing explorers and scientists associated with the North Pole or Antarctica. Talk about those adventurous people who trekked across snow and ice to reach the North Pole or about scientists who live in Antarctica to study the natural phenomena and resources of the area.

Show some maps or pictures of the polar ice caps. Globes and maps will give students some notion of the relative sizes of the ice caps. Photographs from NASA Landsat satellites show the Arctic and Antarctic regions in a dramatic way. These images can be used to initiate a discussion of what measurements of ice masses are important for scientists to make.

Allow students to brainstorm in small groups to come up with ideas about what scientists would want to know about ice that occurs in nature in such large quantities. After the brainstorming session, allow the groups to share one or two ideas each. Make a class list of the attributes of ice masses that would interest scientists.

Lead a discussion about which attributes on the list could be measured using instruments on satellites. Although students' knowledge of sensing instrumentation will vary, all ideas should be accepted. Most students will know that photographs are taken from satellites and that these images can be compared over time. Comparisons of pictures from the same seasons of the year can reveal the existence of perceptible differences in the area of the ice caps.

Developing the Activity

Focus the discussion on measuring the area of an ice mass. Students will need to consider how to measure the area of irregular figures. Have

A sample class list of attributes of an ice sheet

Length
Volume
Width
Temperature
Acidity
Depth
Weight/mass
Surface area
Composition

Teaching Note

For the first method of approximating the area, copy appendix page A-25 and have students lay the page on top of a piece of graph paper. They should be able to see the graph-paper squares through the copy.

Teaching Note

Two methods of estimating the area of the partial squares are suggested. After students try the two methods, have them discuss which method would be better to use with a calculator and which, without a calculator. The "one-half method" is easily accomplished mentally by counting the partial squares and dividing the result by 2. Estimating more precisely requires adding fractions, a method more suited to using fraction calculators. Making decisions about when to use technology is an important part of mathematics today.

them return to their groups and brainstorm ideas for doing so. Once a group reports that it has devised a method for approximating the area of irregular figures, give members copies of the figure on page A-25 to try their technique by determining the area of the ice mass represented by the scaled figure.

After the groups have used their techniques to approximate the area of the irregular ice mass, have them share their procedures and answers with the class. Encourage students to try the methods shared by other groups.

Explain the following three techniques to students if the groups do not propose them. Have students try each technique.

1. Trace the figure onto graph paper, and count the squares inside the figure. Of course, the issue of partial squares is always a topic of conversation. Let the students use their knowledge of fractions to estimate the area of each partial piece. These areas can be added using paper-and-pencil algorithms or a fraction calculator. Another method for estimating the area of partial squares is to count each of them as a half regardless of its size. Have students compare the results of using this method with the approximations made by using more "precise" fractions.
2. Draw the largest possible rectangle inside the irregular area. This rectangle should not extend outside any portion of the figure. Record the area of this rectangle. Next draw two rectangles inside another copy of the figure that fill more of it than the single rectangle did. Compute the areas and the sum of the two areas as an approximation of the area of the irregular figure. Repeat the process with three, four, five, and perhaps, six rectangles. As the students compute the sums of these rectangular areas, they should see that the approximation gets closer to the actual area with each increase

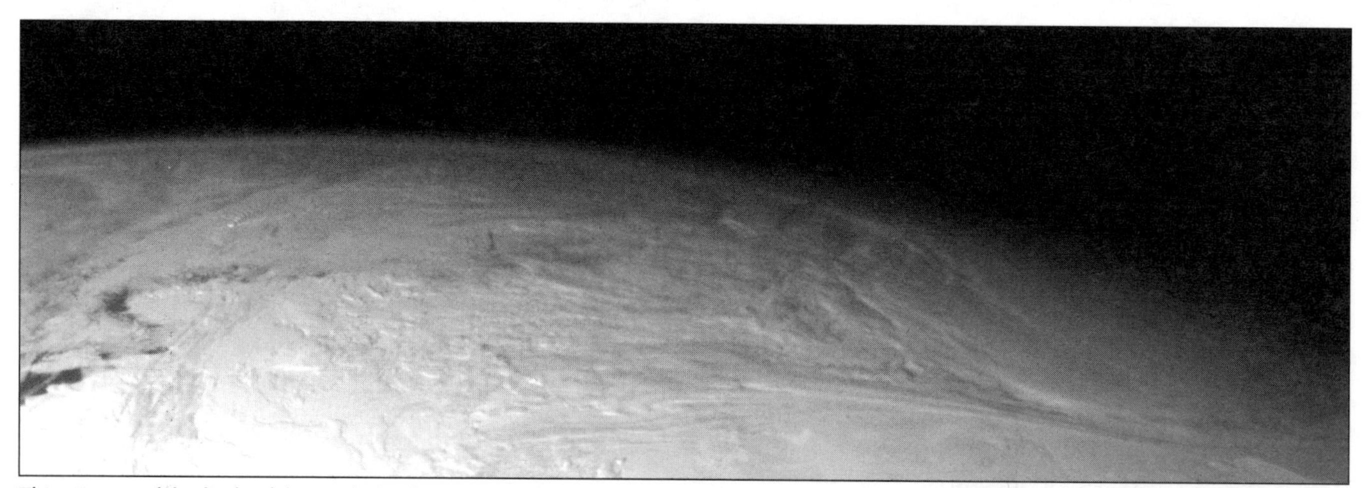

This picture of the limb of the Earth, looking north past Antarctica, is a mosaic of 11 images taken on Dec. 8, 1990, by Galileo's imaging system. The picture spans about 1,600 miles across the south polar latitudes of our planet. The morning day/night terminator is toward the right. The South Pole is out of sight below the picture; the visible areas of Antarctica are those lying generally south of South America. At lower left, the dark Amundsen Sea lies to the left of the Walgreen and Bakutis Coasts. Beyond it, Peter Island reacts with the winds to produce a striking pattern of atmospheric waves.

in the number of rectangles. Let students continue the process until they tire of the task and think they have a reasonable approximation of the area.

3. Use a similar process with external rectangles, and approach the area from an approximation that is greater than the actual area. Draw a large rectangle around the irregular figure that is tangent to it in at least three places. Compute the area. Repeat the process with two smaller rectangles on another copy of the figure. This time, one shared side of each rectangle will pass through the figure so that each rectangle encloses part of the figure. The entire figure should be inside the two rectangles. Repeat the process with an increasing number of rectangles until students realize that the approximation is approaching the value of the actual area.

Concluding the Activity

Students have had several experiences with approximating the area of an irregular figure. Have them write about their experiences in their mission logbooks. Tell them to ask questions about any ideas they did not understand and to specify topics they would like to study again or in more depth. Encourage them to think about the methods of approximation and describe one as if they were explaining the method to a friend.

Also, have students reflect on how NASA scientists might use these same approximation techniques to determine the area of an ice region.

Teaching Note

Some natural phenomena and resources of Antarctica that may interest students include the following:

Penguins
Whales
Seals
Fossils
The size of the ice mass
Temperature ranges
The amount of snowfall
Ozone depletion
Petroleum and natural gas supplies

ACTIVITY 4

Migration of Whales

NCTM Standards

Instructional programs from prekindergarten through grade 12 should enable all students to—

Algebra

Represent and analyze mathematical situations and structures using algebraic symbols

Analyze change in various contexts

Geometry

Use visualization, spatial reasoning, and geometric modeling to solve problems

Measurement

Apply appropriate techniques, tools, and formulas to determine measurements

For many years, scientists and environmentalists have been concerned about the number of whales remaining in the world. This concern continues today. Scientists are studying species of whales to determine, among many things, whether the population of each of the whale species is increasing or decreasing. The use of satellites and radio transmitters attached to selected animals aids scientists in tracking them. In this way, the habits of whales, including migration, can be determined, and scientists can monitor the population dynamics of the mammals.

This activity uses the interdisciplinary nature of the study of whales to apply several mathematical ideas. First, it promotes the use of technology in the form of calculators. Further, the activity introduces the newer technology of global positioning systems (GPS). Handheld GPS units are now available in most stores that sell hunting and fishing gear. The fishing industry uses GPS units to mark locations where fish are plentiful. Hunters and hikers use these same units to locate hunting sites and to find their way from a wilderness location back to a base camp. A great deal of mathematics can be learned from exploring the use of GPS. *Mission Mathematics II: Grades 9–12* (House and Day 2005) discusses the mathematics underlying the calculations accomplished by a GPS.

Many students have some knowledge of whales. Begin this activity with a discussion of what the class knows about these large aquatic mammals. Record all the information offered by students on chart paper, the chalkboard, or a blank transparency. Because this activity focuses on the migration of whales, students should indicate some knowledge of migratory behavior. If students do not seem to have much knowledge about whales, particularly the migration of whales, this activity becomes an opportunity for them to do research. Encyclopedias, science textbooks, and library books all have information about whales. Students with access to the Internet can consult Web sites that focus on whales. Use one of the Internet search engines to find these sites. This research

A right whale breaching

may require some time to complete unless the materials have been collected before the activity. After the class completes its research, allow students to add new findings to the original list of information about whales.

Whale watching is a common tourist activity in the state of Maine. One of the intriguing things about whale watching is that individual whales have identifying features that allow avid whale watchers to identify their favorite animals. These identifying characteristics—special coloring, distinctive fins, and so on—are useful to scientists, too. The progress of an individual whale can be tracked from sightings along its migration path.

Important Mathematical Ideas

Students make accurate measurements with a ruler or yardstick and determine distances by using ratios from scaled maps.

Mission

This activity gives students experience with real-world applications of mathematics from the middle school curriculum. Technology is available that allows people to use mathematics in many settings. This activity involves the use of calculators and, if available, GPS units to explore mathematical ideas associated with distances and locations on Earth.

Materials and Equipment

Rulers, maps of the eastern coast of the United States that include the West Indies, and log sheets (see appendix, page A-2)

Launching the Activity

Humpback whales spend April through November in the nutrient-rich waters of the northern Atlantic Ocean. The whales spend these months feeding and adding many layers of blubber. These layers of blubber provide the adult whale with nourishment for the round-trip journey between the North Atlantic and the Caribbean, where baby whales, or calves, are born. The mothers do not feed while in the quiet, warm waters of the Caribbean. The calves feed on their mother's nourishing milk to prepare them for the return journey to the North Atlantic. In about March, the whales begin the trip home.

This activity can be accomplished in a large group by using a class map. If several maps are available, the activity can be done in small groups.

Making Connections

The activities involving whales have natural connections with both science and social studies. Some ideas about the nature of whales are clearly part of the science curriculum, and the migration of whales focuses attention on geography.

The TOPEX/Poseidon satellite before its launch in September1992

Teaching Tip

This activity uses the migration of humpback whales on the East Coast of the United States as a context. For students who live nearer the Pacific Ocean, studying the gray whale may be more appropriate and motivating. With a minimum of research, information about gray whales can be found in school libraries. Important data for teachers to have for the activity are the range of migration and the names of some geographic locations at each end of the range. Students will add most of the other data for the activity from their research.

Teaching Note

This activity simplifies the migration path of whales. Whales actually swim in a meandering path influenced by currents in the ocean, the availability of food, and the topography of the ocean floor.

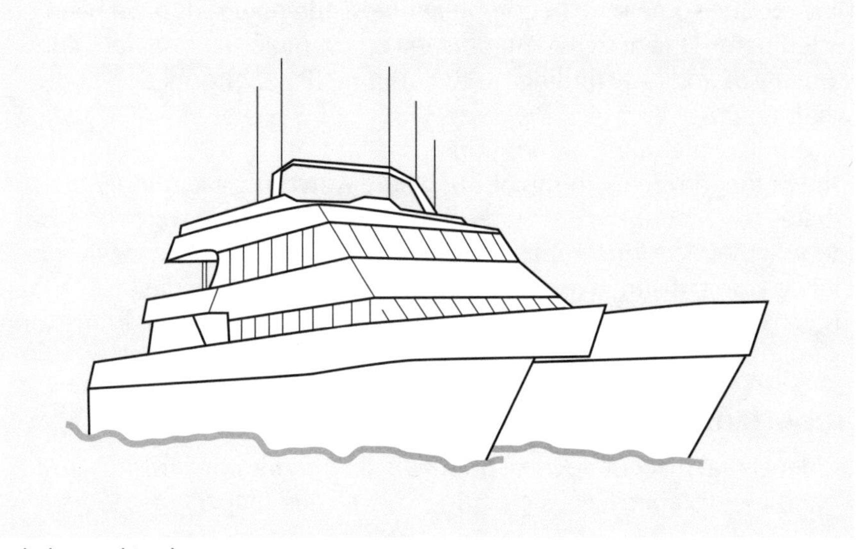

A whale-watching boat

If multiple copies are made from a master map (see appendix, page A-26), students can work individually, as well.

Even in a large group, each student needs to make map measurements, record data, and perform calculations. For example, if a large classroom map is used for the activity, prepare the map for the activity by placing self-sticking, peel-off numbered dots in the ocean along the East Coast of the United States. Use enough dots, representing sightings, to trace a path for the migration from Maine to Florida and from Florida to the West Indies. To allow each student in the class to make a measurement on the class map, trace a return migration path with dots of a different color.

If each student can be given a map of the path of the whale, plot the migration path on a master map and make a copy for each student. Include enough data points that the migration path can easily be imagined and sketched. Keep in mind, however, that too many data points will make the activity tedious for students. The path should probably be limited to ten or fewer data points. Because each student will make many measurements, the teacher can decide whether a different migration path is needed for the return trip. If the same data points are used under the assumption that the whale returns on a similar route, the one-way trip can be doubled to find the total distance traveled.

Have students calculate the distances between the data points on the map. Of course, the map selected will determine the scale to be used. The following solution path shows the calculation of the distance between two points on a map with a scale of 1 inch = 116 miles.

The distance between the first two sightings of the whale in the Gulf of Maine was $2\frac{3}{4}$ inches. How far had the whale traveled?

$$\frac{116 \text{ mi}}{1 \text{ in}} \times \frac{2\frac{3}{4}}{1} \text{ in} = \frac{116 \text{ mi}}{1 \text{ in}} \times \frac{\frac{11}{4}}{1} \text{ in}$$

$$= \frac{1{,}276 \text{ mi}}{4}$$

$$= 319 \text{ mi}$$

Developing the Activity

Complete all the measurements and calculations of the distances between proximate data points on the migration path. Calculate the sum of all the distances to determine the one-way distance that humpback whales migrate from the North Atlantic to the Caribbean.

Humpback whales usually complete this journey in one to two months. Of course, whales do not travel at a constant speed. However, if we know the approximate period of time required to complete the journey, an approximation of the average speed of a migrating humpback can be determined.

Assume that the trip from the Gulf of Maine to the West Indies takes 30 days. What is the average speed of the whale?

$$30 \text{ days} = 720 \text{ h}$$

If the calculated distance is 3,960 miles, then

$$\text{distance} = \text{rate} \times \text{time},$$

or

$$\text{rate} = \frac{\text{distance}}{\text{time}}.$$

The average speed of the whale is

$$\frac{3960 \text{ mi}}{720 \text{ h}}.$$

By dividing, we find that the average speed of the whale is 5.5 miles per hour.

Have students calculate the average speed for the whole journey if the whale takes thirty-six days to complete the one-way trip.

Concluding the Activity

Students should fill out mission log sheets for this activity. Have them attach their papers showing the computations of the distances between points along the migration path and the total distance, along with the computation of the average migration speed of the whales.

Technology Tip

Given that the scale of the map is constant and the problem calls for repeated calculations, this activity presents an opportunity to teach students to perform constant calculations and to use the memory of the calculator to record partial sums of the distances.

Using Calculators

This problem requires long division with multidigit divisors. Using a calculator will allow all students to complete the problem. Having facility with handwritten calculation is not a prerequisite to developing an understanding of dealing with equations and average rates of speed.

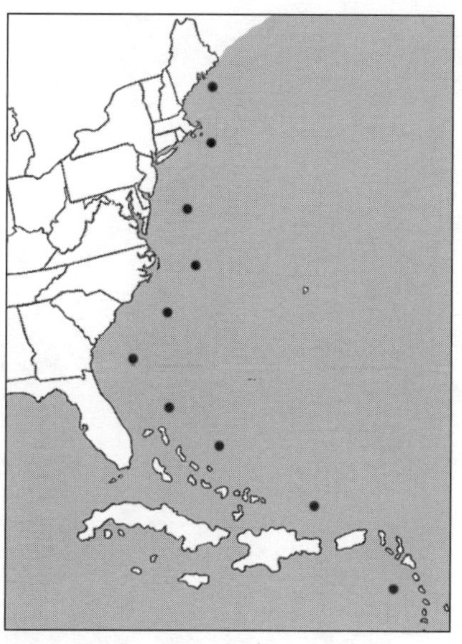

Data points on a map represent a whale's migration route.

ACTIVITY 5

Finding Whales

NCTM Standards

Instructional programs from prekindergarten through grade 12 should enable all students to—

Geometry

Analyze characteristics and properties of two- and three-dimensional geometric shapes and develop arguments about geometric relationships

Specify locations and describe spatial relationships using coordinate geometry and other representational systems

Algebra

Represent and analyze mathematical situations and structures using algebraic symbols

Important Mathematical Ideas

This activity focuses on the use of the Pythagorean theorem to calculate the approximate distance that whales may have traveled. A story about whale watching is used to stimulate the problem solving associated with finding the distance that whales have traveled in a period of time.

Mission

Students locate whales using the Pythagorean theorem to determine distances; they also read a directional compass.

Materials and Equipment

A copy of the worksheet "Thar They Blow" (see appendix, page A-27) for each student, a directional compass.

Many classes may include at least one Boy Scout or Girl Scout, and one of these students may be able to make a presentation about the compass. The leisure activity of orienteering involves the use of a compass. Orienteering compasses, as well as basic ones, are available from outdoor stores at reasonable prices.

Launching the Activity

Ask the class if anyone has had a chance to go whale watching. In most classes, students will not have had this opportunity, and the teacher will have to describe the experience of going out on boats to see whales. Show the class some pictures of humpback whales. Discuss how different humpback whales can be identified by distinctive coloring or by scars on the flukes of their huge tails or on their side flippers.

The student worksheet for this activity uses a fictitious story about a sea captain who operates a whale-watching excursion boat. The purpose of the story is to motivate the students to calculate the approximate distance that whales travel in a short time. Some students will realize that whales do not travel in straight lines. They meander through the water, and the topology of the sea bottom, as well as the availability of food, can affect their paths. However, using straight lines allows middle school students to explore the use of the Pythagorean theorem.

Developing the Activity

Hand out a copy of "Thar They Blow" on page A-27 of the appendix to each student. Have students read the story, and encourage them to draw pictures corresponding to the story. Remind students that on a conventional map, north is at the top, east is to the right, south is at the bottom, and west is to the left. A sample drawing is shown on the facing page.

The example uses a Pythagorean triple (3, 4, 5). Another triple is used to describe a different search in the student worksheet for this activity. If students are ready to apply the Pythagorean equation $c^2 = a^2 + b^2$, where *a, b,* and *c* are rational numbers, the use of calculators is warranted. When designing new problems for the students to consider, be certain to use directions that involve right angles. Also, because the directions of the compass are important in this activity, students should draw pictures with the orientation of the triangle dictated by the directions given.

Concluding the Activity

Students should complete enough teacher-made examples to demonstrate their understanding of—

- how to compute the distance along the hypotenuse of a right triangle when the distances along the legs of the triangle are known and
- how to orient the described triangles according to the given compass directions.

Students' log sheets for this activity should give evidence of their understanding of both these concepts.

Some students will want to know more about how Captain Ahab knew that he had traveled a certain distance on open water. A succinct answer is that the computer chip in the loran (long-range navigation) receiver onboard the boat does that calculation. GPS technology also uses readings from satellites to determine such distances. In tracking whales by satellite, transmitters are attached to whales, and the satellites can "take a reading" of their position as they surface.

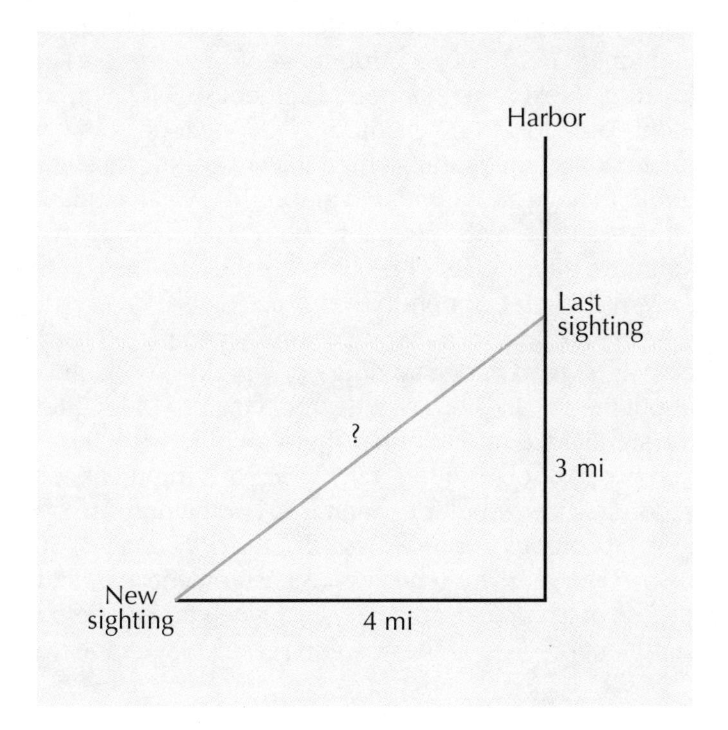

Technology Information

A long-range navigation (loran) unit is a device that provides latitude and longitude information. This land-based system uses radio signals to help sailors, both professional and recreational, determine their locations, keep logs of important locations, and chart journeys from one point to another.

ACTIVITY 6

GPS

NCTM Standards

Instructional programs from prekindergarten through grade 12 should enable all students to—

Number and Operations

Understand numbers, ways of representing numbers, relationships among numbers, and number systems

Geometry

Specify locations and describe spatial relationships using coordinate geometry and other representational systems

Map-Reading Vocabulary

Lines of latitude, called *parallels,* are drawn in an east-west direction and provide a measure, in degrees, of the distance north or south of the equator. The equator has a latitude of 0 degrees, and the North Pole has a latitude of 90 degrees. Lines of longitude, called *meridians,* are the north-south lines on a globe that measure the distance east or west of an arbitrarily selected meridian. This 0 degree meridian is known as the *prime meridian.* It runs from the North Pole to the South Pole through Greenwich, a town in England.

This activity helps develop understanding of two concepts. The ability to determine a location on the globe from latitude and longitude readings is a prerequisite to thinking about the bearing from one location to another and the track, or path, of someone moving between the two locations.

Mission

Students experience mathematics using the terminology of geography.

Materials and Equipment

A compass; a globe; a paper plate, a brass fastener, and two strips of index-card stock for each student; a handheld GPS unit (optional)

Student sheets for this activity can be found in the appendix on pages A-28–A-32.

Launching the Activity

Open the activity by showing the class a globe. Have students tell what they know about the globe, such as the number of continents, names of oceans, locations of countries and hemispheres, and so on. In some classes, a student may mention the lines of longitude and latitude. If no one points out these lines, ask students about them.

After the class has discussed longitude and latitude, give students practice in using these concepts. Spend some time having students come to the globe in pairs. Ask one student to locate a given latitude and longitude. The other student confirms or refutes the selection of the location. Next have the second student specify a given city or other identifiable location by latitude and longitude, and have the first student confirm the reading. Time may not allow all students in a large group to have a turn with the globe. To give all students experience with these concepts, put the globe or a world map in a learning center and have pairs of students complete the activities on the worksheet on page A-28 of the appendix.

After all students have successfully completed the learning-center activity, extend understanding of longitude and latitude by talking about the subdivisions of a degree. Degrees of longitude and latitude are subdivided into minutes, the symbol for which is ′. Each degree is composed of 60 minutes. Likewise, each minute is subdivided into 60 seconds, the symbol for which is ″. For the purpose of reading a map or globe, positions as precise as 38°28′36″ N and 78°50′08″ W cannot be used for a location because measurements cannot be made with that degree of accuracy. However, GPS satellites can help improve measurement precision. The position screen of a handheld GPS unit displays degrees and minutes, along with fractions of minutes that are shown as hundredths of a minute.

Developing the Activity

If a GPS unit can be located to use for a demonstration, show the position screen of the device. A blackline master of the picture at right is provided on page A-29 of the appendix. Note that the design of GPS screens may differ slightly on different brands of receivers, but the information will be the same.

The position screen offers more information than the class is ready to use at this time. Start with the data at the bottom and work upward. Military time is used, although on most GPS units, the time can be changed to the customary format. The altitude shows the height above sea level. Next the longitude and latitude of the location of the receiver are shown. The top part of the screen shows data about the movement of the receiver. The person carrying the GPS causes this movement, whether running, sitting in a moving boat or car, or riding a bicycle. Finally, the name for the location is given.

GPS units have memory, in which the user can store important locations. These stored locations can be used to plan trips, to get from one place on a trail to another, or to locate a favorite "fishin' hole." Scientists aboard a research vessel who learn that migrating whales are located at a specific longitude and latitude can plot a course to catch up with the whales. The screens on the GPS units provide data for these types of activities.

To use the navigational screens effectively, students must be able to read a directional compass, including the degrees on the compass. Most students will be familiar with the directions on a compass, and they have seen many maps in books. They realize that customarily, north is "up" and south is "down." East is to the right of the north-south line, and west is to the left. However, many students may be unaware that each of these directions is measured in degrees. Show the class a directional compass or a transparency of a simulated compass. (A blackline master is furnished on page A-30 of the appendix.)

As can be seen from the drawing, north is 0 degrees, east is 90 degrees, south is 180 degrees, and west is 270 degrees. Have students make conjectures about the measures associated with the other lines in the drawing. Be certain they understand that as an imaginary clock hand approaches north in a clockwise movement, the rotation of the clock hand gets nearer and nearer to 360 degrees, the number of degrees in a circle.

Have each student make a paper-plate compass, similar to a paper-plate clock. The plate forms the compass, and two "hands" are attached. The hands will be added in sequence as the concepts of bearing and track are developed. Students should cut one long, narrow arrow, or

Obtaining GPS Navigators

GPS units are available in sports stores that sell hunting, fishing, and hiking equipment. Parents who enjoy these activities might share their equipment for classroom use, or they may be willing to demonstrate the units. Some sports-store employees may give demonstrations, too.

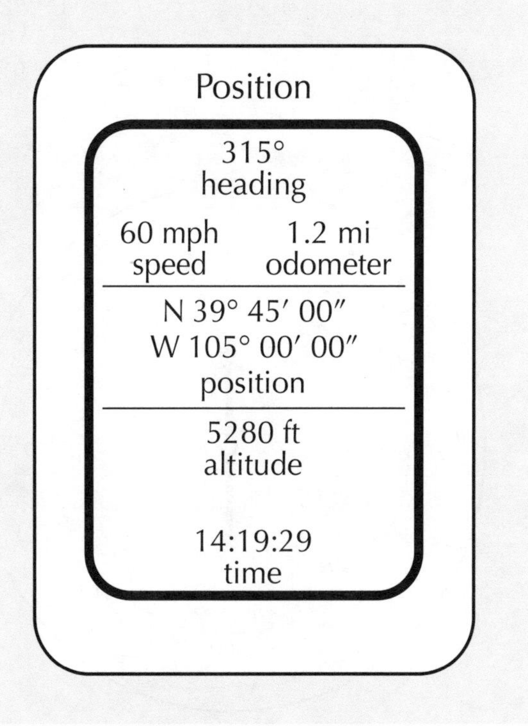

A GPS position screen

Navigation Vocabulary:

Bearing: the intended direction that a person must travel to reach a specified destination

Track, or *course:* the actual direction that a person is traveling

Curriculum Connection

Ask students, "Why is Denver known as the Mile-High City?"

needle, from the index-card stock and attach it to the middle of the paper plate with a brass fastener. Students should then label the degree measures at the appropriate points on the compass for 0°, 45°, 90°, 135°, 180°, 225°, 270°, and 315°.

After students have constructed their paper-plate compasses, they can practice moving the needle to degree settings described by the teacher. When they are able to position the compass needle correctly, have students look at another of the screens—the pointer, or navigation, screen—on the GPS unit or the facsimile on page A-31 of the appendix.

Most information on the pointer screen is similar to that on the position screen. The concepts of bearing and track, or course, are the essential ideas of this activity. Notice that the half compass in the center of the pointer screen has two pointer arrows, one for the bearing and another for the track. Have students add the second arrow to the paper-plate compass. This arrow should be shorter and heavier than the first one. The following example illustrates a use of the bearing and track information given on a pointer screen.

Suppose that a whale researcher using satellite readings has located a pod of whales somewhere in the distance. The latitude and longitude of the pod are saved. Because the objective is to study the whales, the researcher selects the pointer screen. The captain of the research vessel reads the bearing of the pod (293°) to set the track (course) of the ship in that direction. First, the captain notes the bearing of the pod on her or his compass. The next task of the captain is to read the track (course) of the boat and turn the proper direction and amount so that the bearing and track are the same. When the captain looks at the GPS unit, the pointer screen reads as shown on the facing page.

Use the paper-plate compass to show both the bearing and the track for these data. The long hand should show the track, and the shorter one, the bearing, just as the arrows on the pointer screen do. How many degrees is the ship off course? Should the captain turn left or right?

Practice what has been learned about bearing and track using the "Bearings and Tracks" worksheet on page A-32 of the appendix.

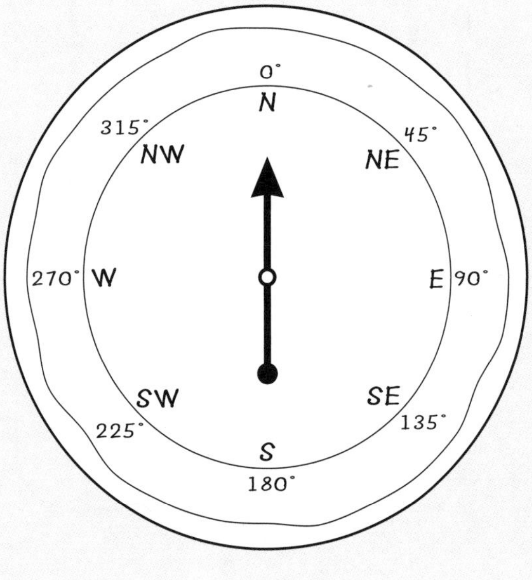

An example of a paper-plate compass

Concluding the Activity

The mission log sheets should be completed to close the activity. Because students have encountered science, geography, and mathematics in this activity, an appropriate culminating task would have them write about all three. The theme of this activity is ideal for an interdisciplinary unit. As a part of the language arts portion of the interdisciplinary unit, students might be asked to develop a public-service announcement about whales and their migration. Technology-minded students might be challenged to design a World Wide Web page to display information about migrating whales.

Students might also be directed to do additional library or Internet research on other scientific activities involved in NASA's Mission to Planet Earth. NASA scientists and laboratories are involved in activities focused on ozone depletion, deforestation, air pollution, ocean temperatures, and many other exciting areas of research. These topics, too, can be the basis for interdisciplinary units in which students learn mathematics in a meaningful and motivational context.

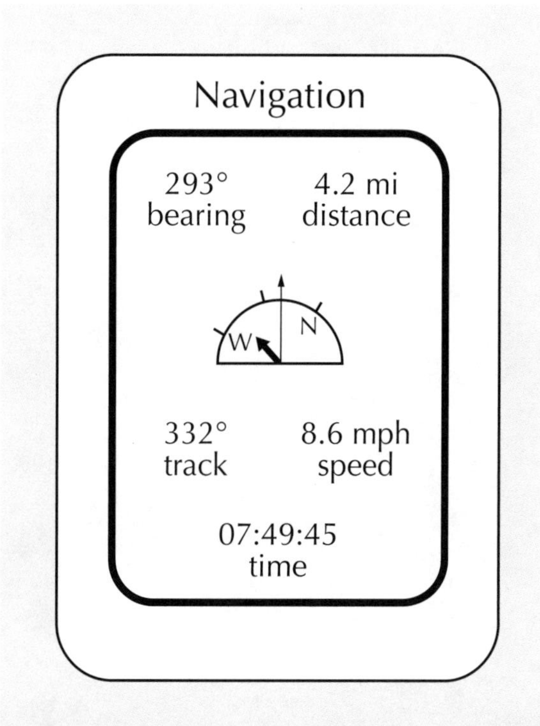

A GPS pointer screen

Students may conclude this module by designing mission patches and logbook covers that relate to environmental concerns, GPS technology, or whales.

Technology Information

The GPS is composed of a constellation of twenty-four Navstar satellites. From unobstructed areas, at least four satellites are "in view." The computer chips in the handheld GPS units read data transmissions for position and elevation from these satellites. A GPS receiver acquires the signal, then measures the interval between the transmission and the receipt of the signal to determine the distance between the satellite and the receiver. This process is called *ranging*. When the receiver calculates the range from at least three satellites, the location of the receiver on Earth can be determined. Handheld GPS devices are accurate enough for hikers, hunters, and fishers. The readings are correct within 100 meters, or approximately 350 feet. More information on calculating the location of GPS receivers can be found in "Finding Our Way" in *Mission Mathematics II: Grades 9–12* (House and Day 2005).

Appendix

STUDENT PAGES

Mission Math Log Sheet

Date: ____________________

Mission Specialist's Name: ____________________

Team Name: ____________________

Members: ____________________

____

____

____

____

The things I did and the mathematics I used:

The things I found out and the mathematics I learned:

The things I wonder about and the mathematics I want to know about:

©2005 by The National Council of Teachers of Mathematics, Inc. www.nctm.org.
All rights reserved. For classroom use only. Written permission required for all other uses.

DATA SHEET

Orbit Simulator

1. Complete the chart showing the relationship between radius and time for thirty revolutions.

Radius (cm)	Time for Thirty Revolutions (s)
85	
80	
70	
65	
60	

2. Plot your data points on graph paper. Label the axes as shown below. Draw a smooth curve through the points. **Warning: unless the points are clearly in a straight line, do not draw a *straight line* through them.**

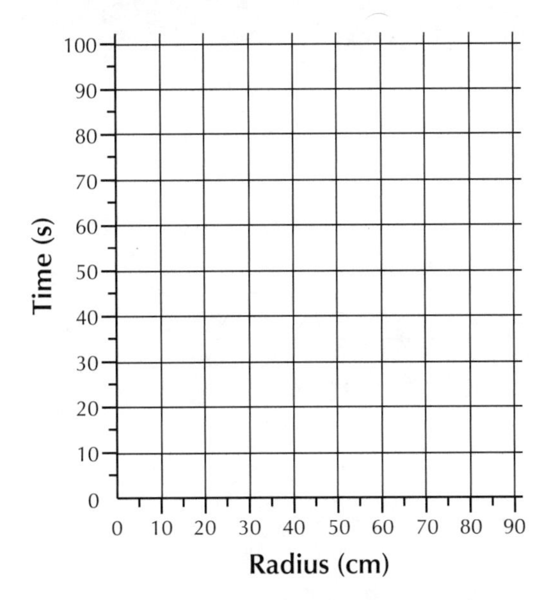

The relationship between radius and time for thirty revolutions

3. Use your data to make the following predictions:

   - If the radius is 75 cm, use your graph to predict the time for thirty revolutions.

     Mark this point on your graph with an *x*, and record your estimate for the time for thirty revolutions:

     ____________ seconds

   - How close to the actual time should you be for the prediction to be good?

     ____

   Explain your answer.

   ____

   ____

   ____

4. Select either the mean, median, or mode to report the "average" predicted time for your class. Record that "average" here:

   Circle one: mean, median, mode = ________

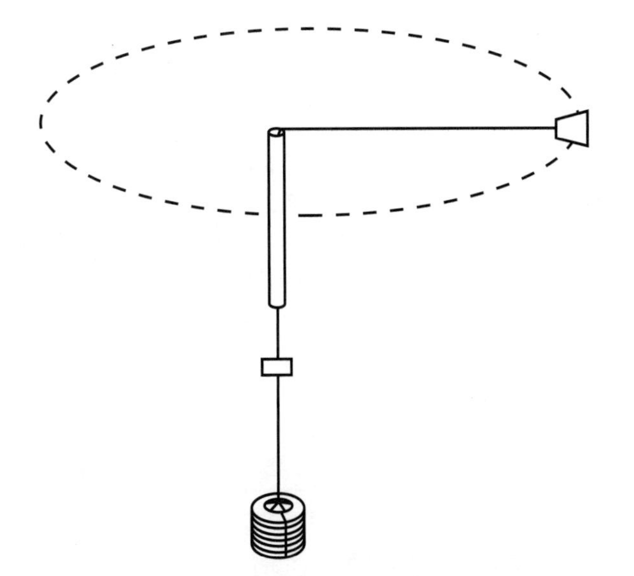

©2005 by The National Council of Teachers of Mathematics, Inc. www.nctm.org.
All rights reserved. For classroom use only. Written permission required for all other uses.

DATA SHEET

Planetary Revolution and Alignment in the Delphi System

Orbital Periods of Delphi Planets

Planets	*Data Set 1*	*Data Set 2*
A	2	1
B	3	2
C	4	3
D	5	4

Table 1

Planet	Period in Gollies*

* One golly is represented by ______ seconds.

Table 2

Prediction of Alignment

Planets Aligned	Time until Alignment
A and B	
A, B, and C	
A, B, C, and D	

Table 3

Actual Time of Alignment from Simulation

Planets Aligned	Time until Alignment
A and B	
A, B, and C	
A, B, C, and D	

Put a check in the column in table 4 that represents the number of gollies for each completed orbit.

Table 4

Orbits Completed by Planets in the Delphi System

Planet	Number of Gollies											
	1	2	3	4	5	6	7	8	9	10	11	12

©2005 by The National Council of Teachers of Mathematics, Inc. www.nctm.org.
All rights reserved. For classroom use only. Written permission required for all other uses.

DATA SHEET

Planetary Alignment in the Solar System

Suppose that the four inner planets are in alignment. When will it happen again?

Use a calculator to determine when the planets are in alignment. The chart on the right below shows subproblems to help you find the answer.

The alignment of all the planets would be a rare event. Can you determine how often it would occur?

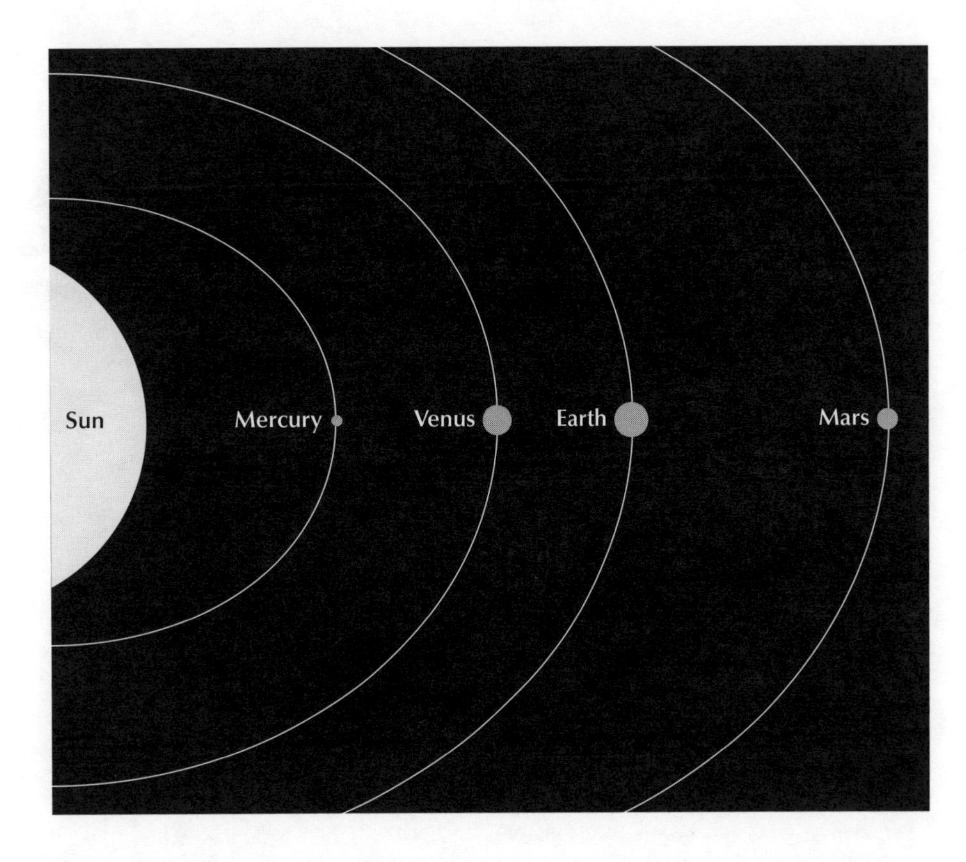

	Planet	Period of Revolution
1	Mercury	88 days
2	Venus	225 days
3	Earth	365 days
4	Mars	687 days
5	Jupiter	12 years
6	Saturn	29 years
7	Uranus	84 years
8	Neptune	165 years
9	Pluto	248 years

Planets Aligned	Years or Days until Alignment
1 and 2	
1, 2, and 3	
1 through 4	
1 through 5	
1 through 6	
1 through 7	
1 through 8	
1 through 9	

©2005 by The National Council of Teachers of Mathematics, Inc. www.nctm.org.
All rights reserved. For classroom use only. Written permission required for all other uses.

The Planets at a Glance

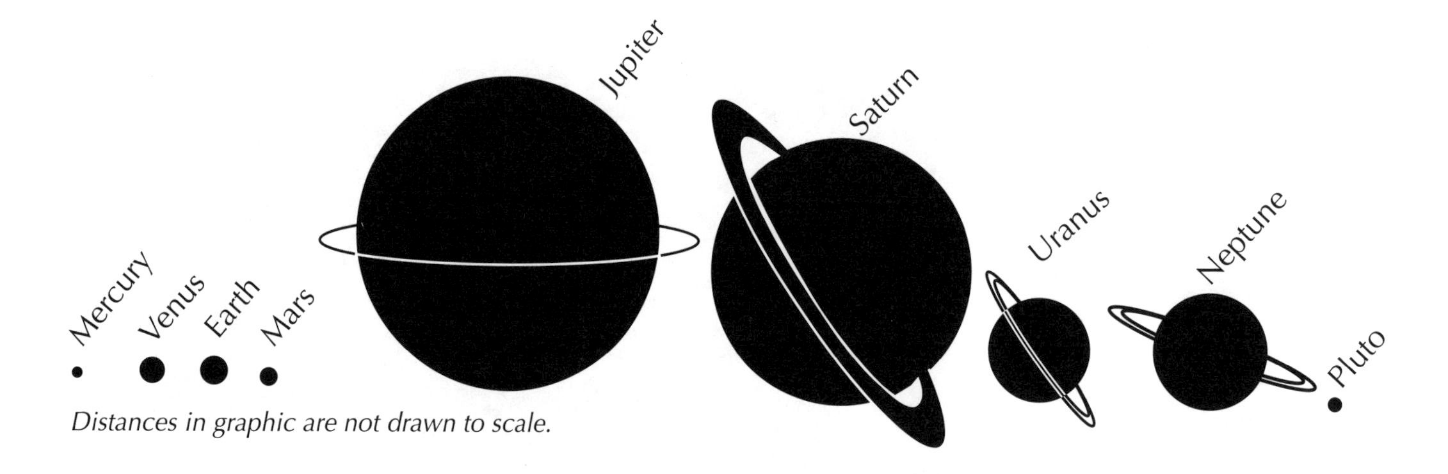

Distances in graphic are not drawn to scale.

	Mercury	Venus	Earth	Mars	Jupiter	Saturn	Uranus	Neptune	Pluto
Mean distance from the Sun									
Millions of kilometers	57.9	108.2	149.6	227.9	778.3	1,427	2,871	4,497	5,913
Millions of miles	36	67.2	93	141.5	483.3	886.2	1,782.9	2,792.6	3,672
Period of revolution	87.9 days	224.7 days	365.3 days	687.0 days	11.86 years	29.46 years	84.0 years	165.0 years	248.0 years
Period of rotation (in Earth time)	58.9 days	243 days retrograde	23 h, 56 min	24 h, 37 min	9 h, 55 min	10 h, 40 min	17 h, 12 min retrograde	16 h, 7 min	6.39 days retrograde
Inclination of axis	0.0°	177.2°	23°27′	25°12′	3°5′	26°4	97°55′	28°48′	120°
Inclination of orbit to ecliptic	7°	3.4°	0°	1.85°	1.3°	2.5°	0.8°	1.8°	17.1°
Eccentricity of orbit	0.206	0.007	0.017	0.093	0.048	0.056	0.046	0.009	0.248
Equatorial diameter									
Kilometers	4,880	12,100	12,756	6,786.8	143,200	120,000	51,800	49,528	~2,330
Miles	3,031	7,514	7,921	4,215	88,927	74,520	32,168	30,757	~1,447
Atmosphere	Essentially none	Carbon dioxide	Nitrogen, Oxygen	Carbon dioxide	Hydrogen, Helium	Hydrogen, Helium	Hydrogen, Helium, Methane	Hydrogen, Helium, Methane	Methane + ?
Number of satellites	0	0	1	2	16	18+ (?)	15	8	1
Number of rings	0	0	0	0	1	1,000 (?)	11	4	0

©2005 by The National Council of Teachers of Mathematics, Inc. www.nctm.org.
All rights reserved. For classroom use only. Written permission required for all other uses.

DATA SHEET

Calculating Proportional Distances for a Model of the Solar System

This ratio sets the scale for the model.

This ratio allows us to find the location of any planet in the model.

$$\frac{\text{Classroom distance from the Sun to Pluto}}{\text{Actual mean distance from the Sun to Pluto}} = \frac{\text{Classroom distance from the Sun to planet}}{\text{Actual mean distance from the Sun to planet}}$$

Planet	Actual Mean Distance from the Sun (millions of km)	Classroom Distance from the Sun	
		First Trial	Second Trial
Mercury	57.9		
Venus	108.2		
Earth	149.6		
Mars	227.9		
Jupiter	778.3		
Saturn	1 427		
Uranus	2 871		
Neptune	4 497		
Pluto	5 913		

©2005 by The National Council of Teachers of Mathematics, Inc. www.nctm.org.
All rights reserved. For classroom use only. Written permission required for all other uses.

DATA SHEET

How Far Can I Go in 8 Seconds?

Complete the charts below. On a single graph of distance versus time, graph the points that would indicate the distance traveled in each mode in 8 seconds. Draw a line from the origin to each point and label the line to indicate the mode of transportation and the speed. If a computer is available, use a spreadsheet to perform your calculations.

Mode of Travel	Distance**	Time**	Distance in Feet	Time in Seconds
Student*				
Bike (flying start)		0.0019 h	656.2	
Car	8 mi	45.5 s		
Rocket-powered three-wheel car	3 mi			17.11
Propeller-driven airplane	1.8 mi	0.0034 h		
Jet airplane	2,982 mi	3.9236 h		
Space Shuttle	4,164,183 mi	9 d, 23 h, 30 min		

* Enter data from your classroom experiments.
** The given data indicate recognized "world records" for each mode of travel.

Mode	Speed (ft/s)	Distance in 8 Seconds
Student		
Bike		
Car		
Rocket-powered three-wheel car		
Propeller-driven airplane		
Jet airplane		
Space Shuttle		

$$\text{Speed} = \frac{\text{Distance}}{\text{Time}}$$

©2005 by The National Council of Teachers of Mathematics, Inc. www.nctm.org.
All rights reserved. For classroom use only. Written permission required for all other uses.

The 757

To assemble your 757, see the instructions on page A-10.

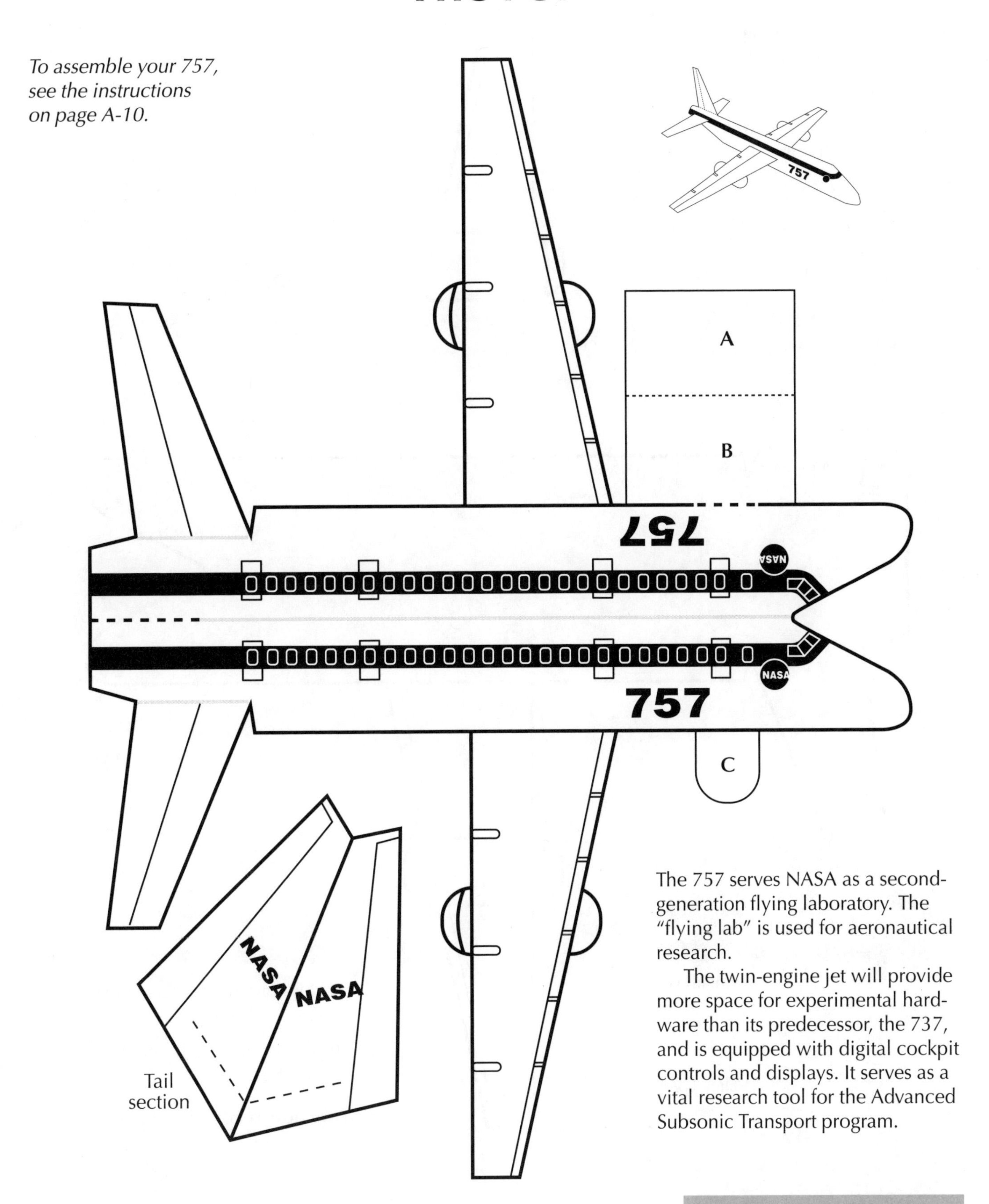

The 757 serves NASA as a second-generation flying laboratory. The "flying lab" is used for aeronautical research.

The twin-engine jet will provide more space for experimental hardware than its predecessor, the 737, and is equipped with digital cockpit controls and displays. It serves as a vital research tool for the Advanced Subsonic Transport program.

©2005 by The National Council of Teachers of Mathematics, Inc. www.nctm.org.
All rights reserved. For classroom use only. Written permission required for all other uses.

Directions for Assembling the 757

1. Cut all the parts on both pages with scissors, and glue together the unmarked sides of the two large pieces.
2. Cut the heavy dashed lines.
3. Mountain-fold* the body of the plane in half lengthwise.
4. Valley-fold** A inward to B. Fold B into the plane section. This step adds weight to the front of the plane.
5. Insert tab C into the slot indicated by the dashed line cut at the bottom of B.
6. Mountain-fold the tail section along the center line, and slide it into the slot in the back of the plane until it locks into place. Secure the tail with tape so that it stays in place.

* Mountain Fold

** Valley Fold

7. Valley-fold the front and back "wings" to form a glider.
8. Tape and paper-clip the nose of the airplane.

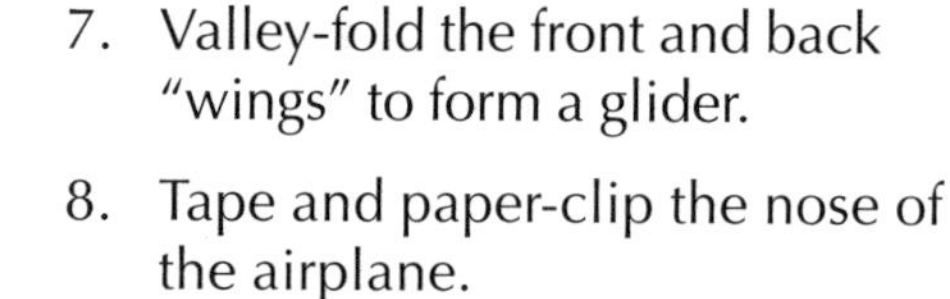

You are now ready to fly your 757!

©2005 by The National Council of Teachers of Mathematics, Inc. www.nctm.org.
All rights reserved. For classroom use only. Written permission required for all other uses.

EXPERT GROUP 1

The Egret

Directions for Constructing the Egret

Step 1

Use a full sheet of typing paper. Using the mountain fold, fold the paper in half lengthwise. Unfold. Then valley-fold the upper corners to the center crease.

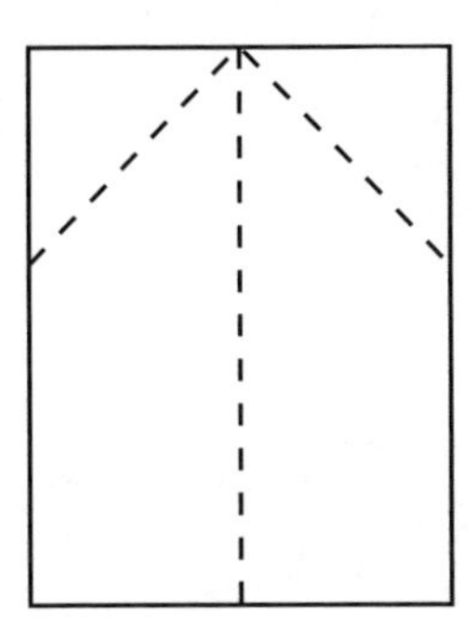

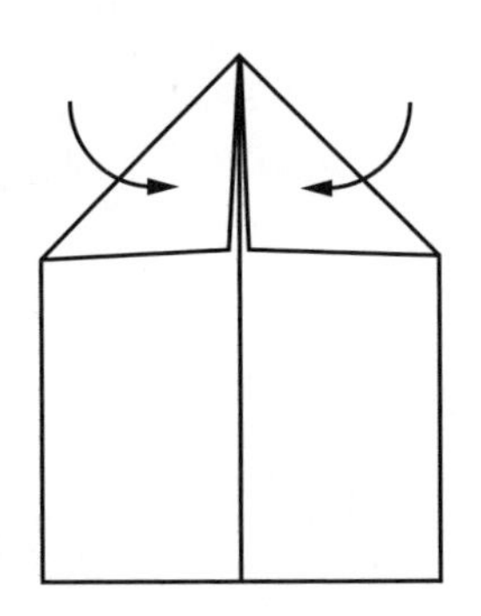

Step 2

Valley-fold along the dashed lines to form triangles that meet the center crease, as shown.

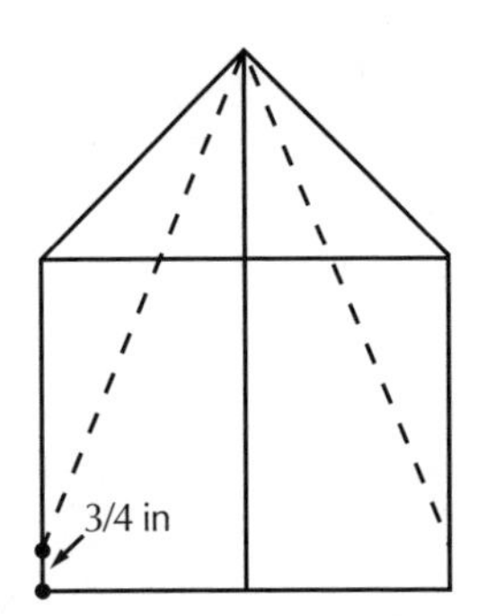

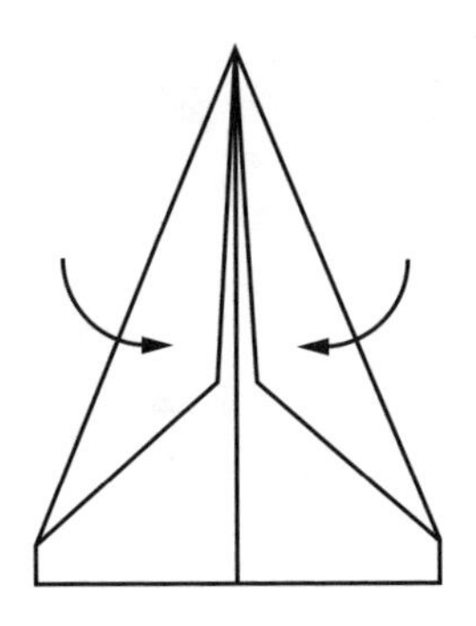

Step 3

Valley-fold the outer edges to meet the center crease as shown by the dashed lines. Unfold as shown.

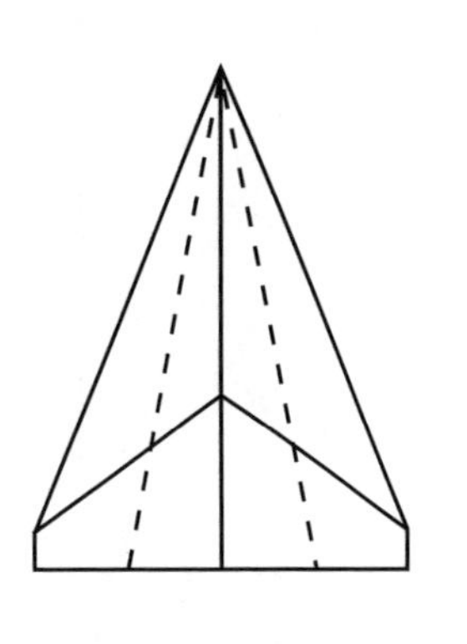

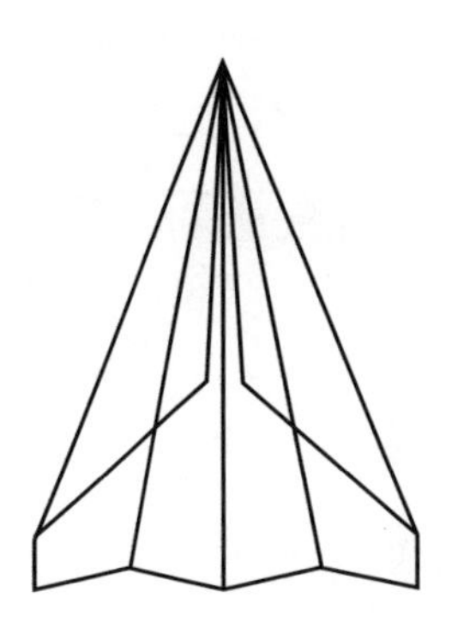

Making the Airplane

The Egret is constructed similarly to the common paper airplane that everyone makes. But because of this model's carefully measured shape, it can attain a very smooth and flat glide. Make sure that its shape is properly adjusted, with vertical tails straight up and down. Hold it between thumb and forefinger, launching it gently straight ahead.

©2005 by The National Council of Teachers of Mathematics, Inc. www.nctm.org.
All rights reserved. For classroom use only. Written permission required for all other uses.

Step 4

On each side, measure along the diagonal edge of the paper, as shown by the heavy line, and cut. Along the bottom edge, measure 5/8 in (1.6 cm) from each wingtip, and from this point, draw a line to the end of the cut. Valley-fold along this line to make vertical tails.

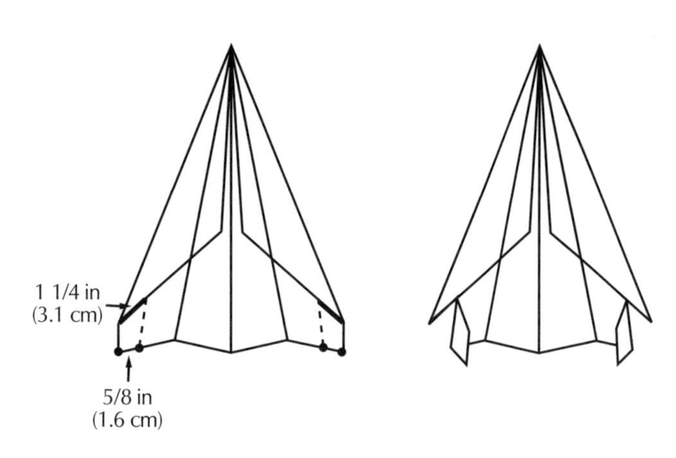

Step 5

In the locations shown, find the center line, measure, cut, and fold the elevators.

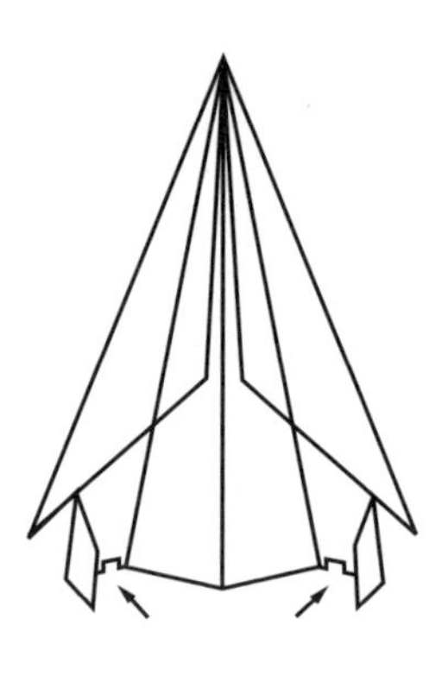

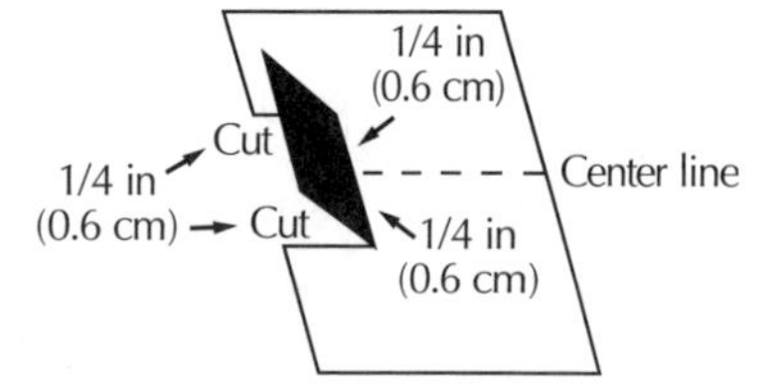

Step 6

Glue folds only at the center of the fuselage. Flip the airplane over. Adjust the shape so that when viewed from the back, the airplane makes a shallow upside-down W, as shown.

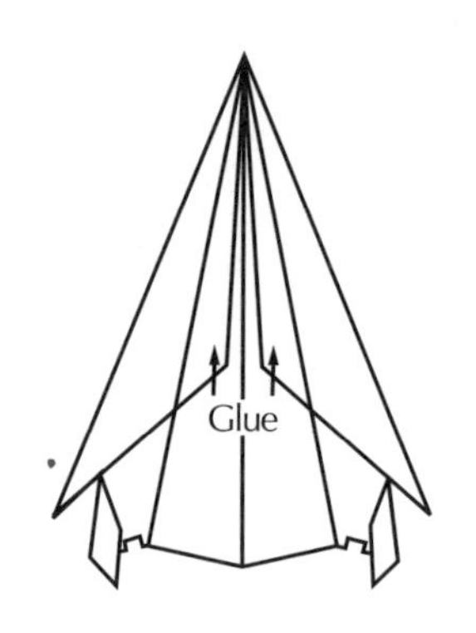

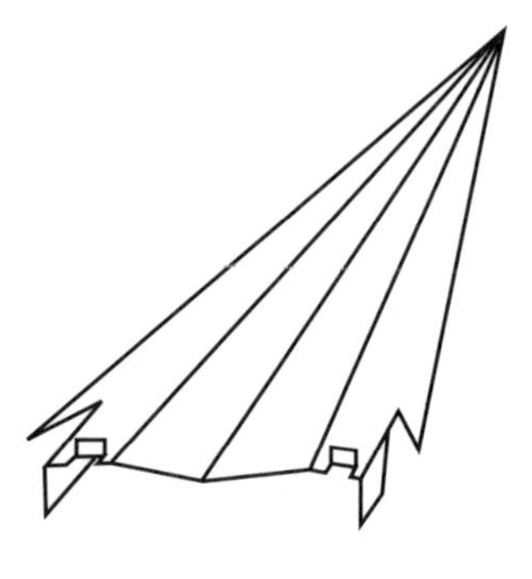

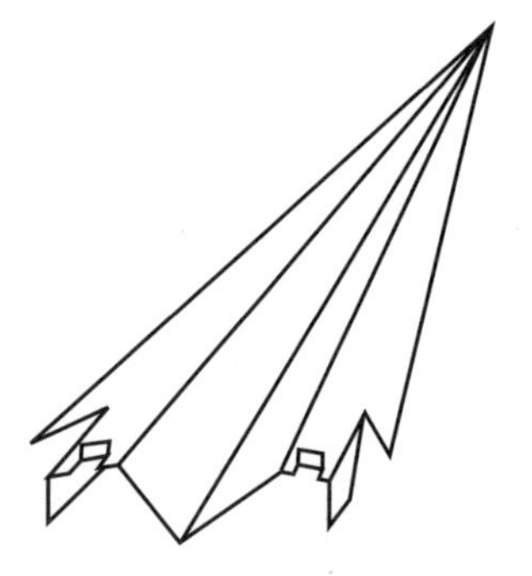

©2005 by The National Council of Teachers of Mathematics, Inc. www.nctm.org.
All rights reserved. For classroom use only. Written permission required for all other uses.

EXPERT GROUP 2

The Flex

Directions for Constructing the Flex

1. Find center line A of a sheet of typing paper. Valley-fold on that line, and crease it. Then open the sheet.
2. Mark off seven lines, each 0.5 inch apart, parallel to the left lead edge. Valley-fold and crease along each line, "rolling" the paper flat toward the tail end.
3. Valley-fold on center line B. Cut out the wings and tail section of the plane as shown on the pattern.
4. Mountain-fold on both sides of line B about 0.5 inch from line B to form the body of the plane.
5. Valley-fold to form the tail flaps, and mountain-fold to form the wing flaps as indicated by dashed lines C and D on the pattern.

By changing the positions of the wing and tail flaps, students can make this plane perform many different maneuvers.

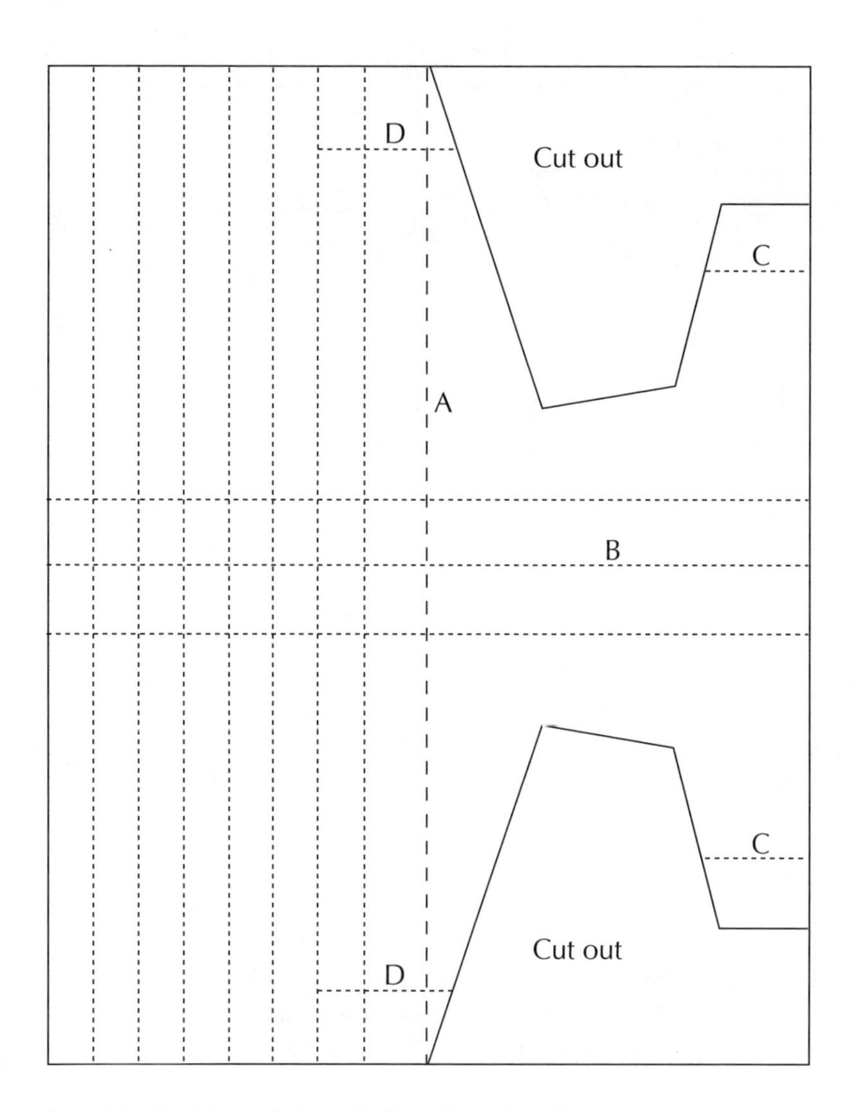

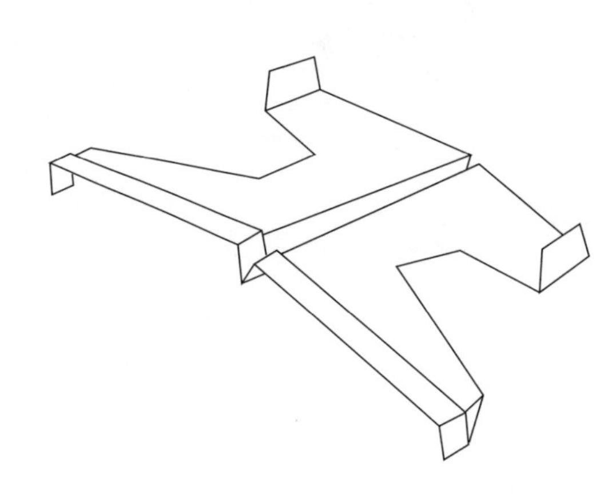

©2005 by The National Council of Teachers of Mathematics, Inc. www.nctm.org.
All rights reserved. For classroom use only. Written permission required for all other uses.

EXPERT GROUP 3

The Basic Square

Directions for Constructing the Basic Square

1. Use a full sheet of typing paper (see fig. 1).
2. Begin to make parallel valley folds, starting at the bottom of the sheet (see fig. 2).
3. Continue until the "wing" areas are just about square (see fig. 3).
4. Valley-fold along the center line (see fig. 4).
5. Mountain-fold the two wings parallel to the center line, leaving about 0.5 inch on either side of the line for the body of the plane (see figs. 5 and 6).
6. Be sure to crease the folds nicely at the front of the plane.

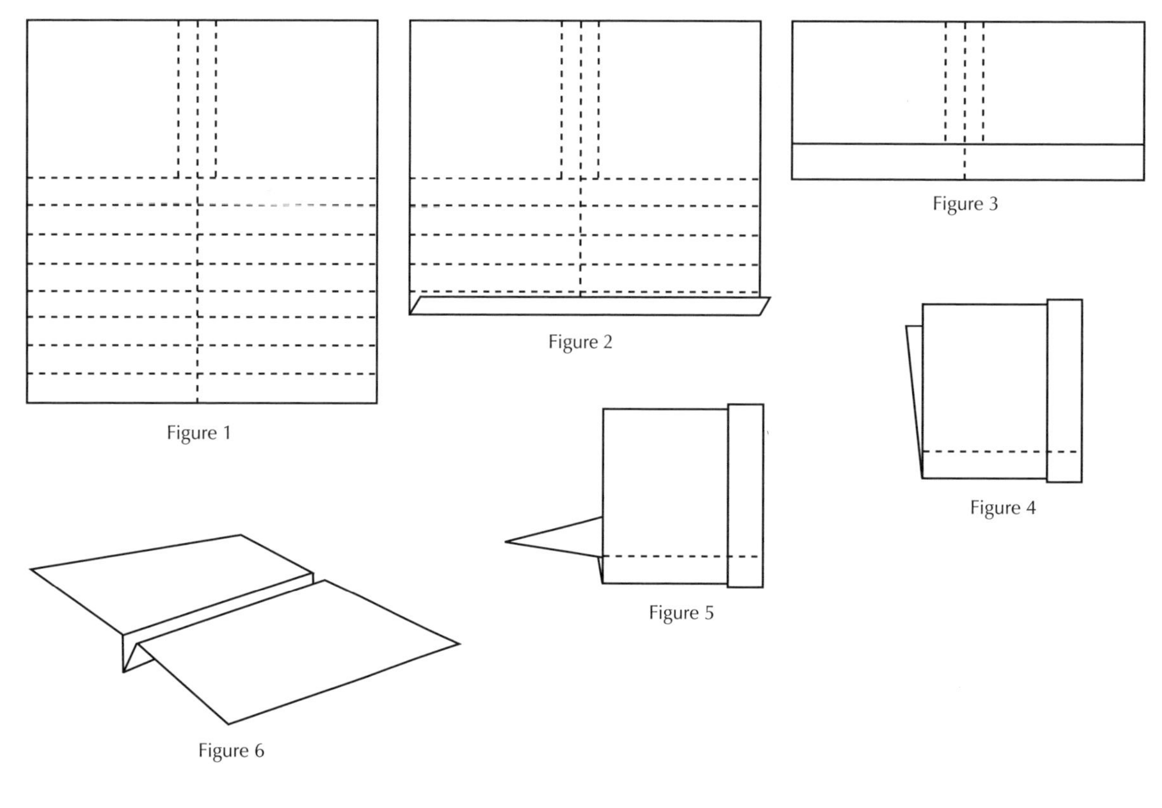

Figure 1

Figure 2

Figure 3

Figure 4

Figure 5

Figure 6

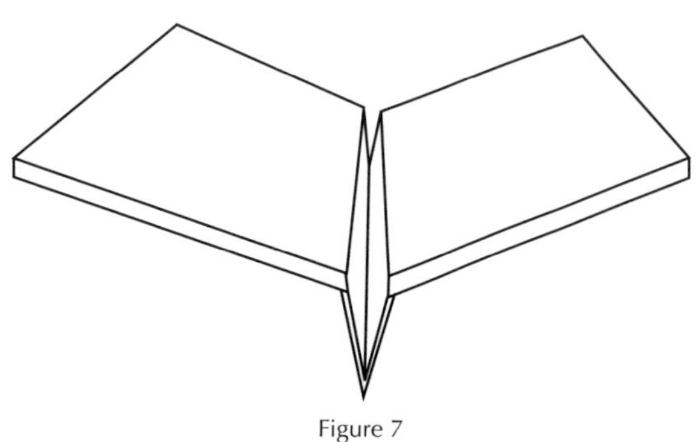

Figure 7

7. To fly well, your plane should have its wings higher than the body so as to form a Y shape (see fig. 7).

This plane is modeled after the "Basic Square," described by Ken Blackburn in *The World Record Paper Airplane Book* (Blackburn and Lammers 1994). Blackburn is the world-record holder for keeping a paper airplane aloft (18.80 seconds).

©2005 by The National Council of Teachers of Mathematics, Inc. www.nctm.org.
All rights reserved. For classroom use only. Written permission required for all other uses.

EXPERT GROUP 4

The Condor

Directions for Constructing the Condor

Step 1

Use a full sheet of typing paper. Mountain-fold the paper in half lengthwise. Unfold. Then valley-fold it in half perpendicular to the first fold. Unfold.

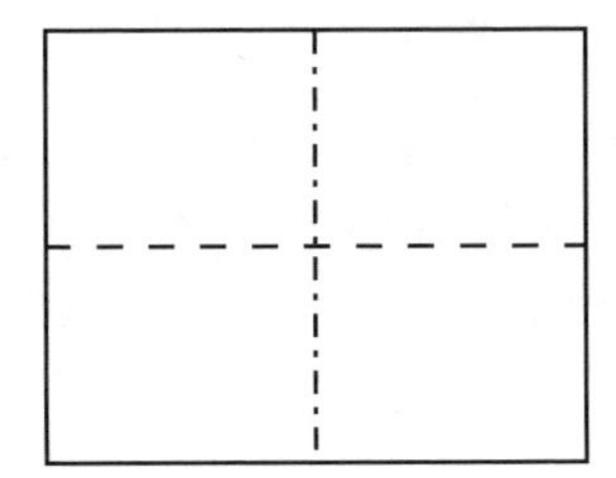

Step 2

On each side, valley-fold the top corners diagonally so that the outer edges meet the horizontal crease.

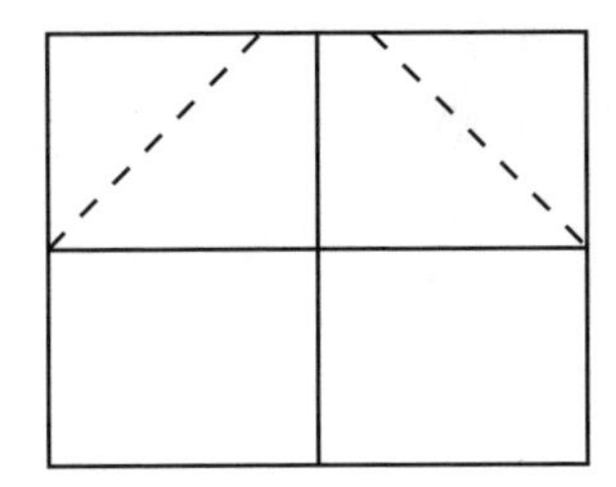

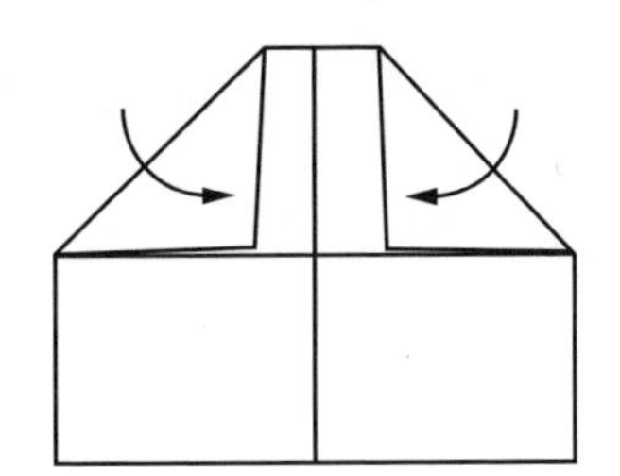

Step 3

Make a valley fold parallel to the step-1 lengthwise crease so that the top edge meets the crease. Make another valley fold so that the top edge meets the crease. Then refold the original lengthwise crease.

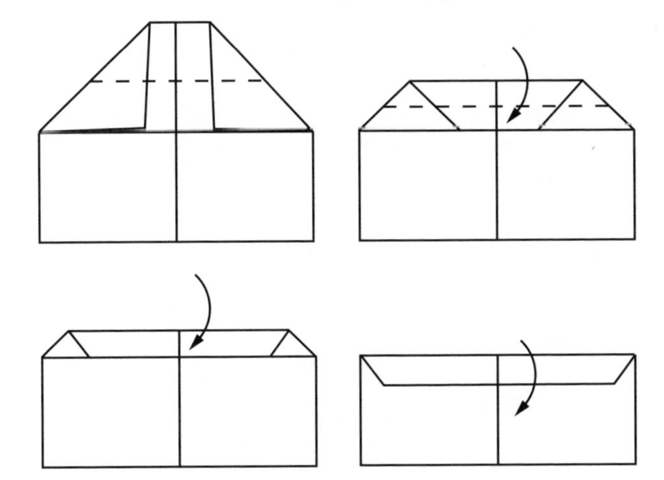

Step 4

On each side of the step-1 perpendicular crease, measure and draw diagonal lines as shown. Valley-fold the top outer edges along these lines. Glue the folded-over triangles to form the leading (front) edges of the wings. Then measure and draw lines as shown. Valley-fold along these lines to form the fuselage.

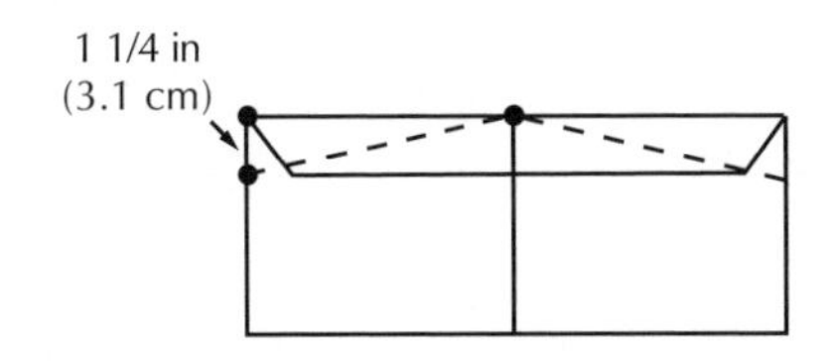

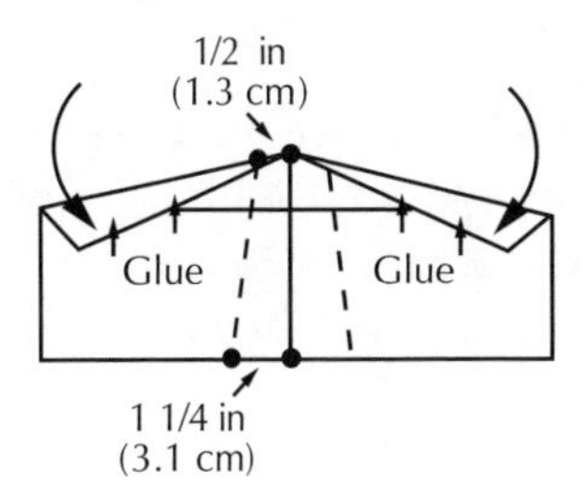

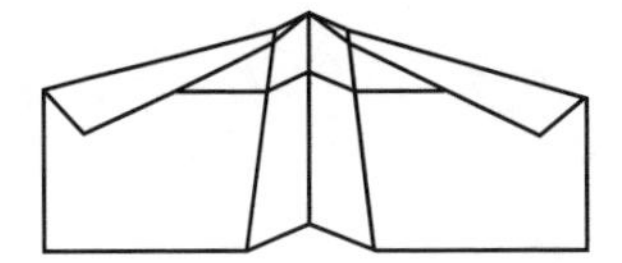

Making the Airplane

Condors have large feathers at their wingtips for control. Instead of feathers, this airplane has winglets. Because of its wide wing span, this paper airplane is fragile where the wings meet the fuselage. Adjust the winglets, and bend the airplane to adjust the trim.

©2005 by The National Council of Teachers of Mathematics, Inc. www.nctm.org.
All rights reserved. For classroom use only. Written permission required for all other uses.

Step 5

Flip the plane over. On each side, measure and draw the lines for the winglets as shown. Cut on the heavy lines. Valley-fold as indicated to make the winglets. Make the canopy.

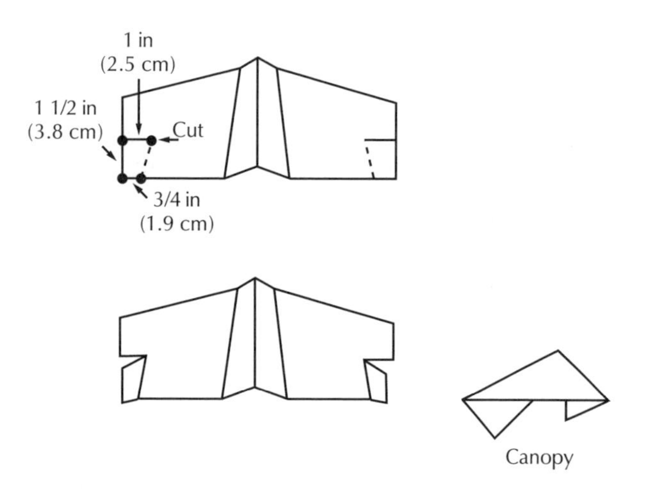

Making the Canopy

1. Measure and cut a 2″ × 3″ (5 cm × 7.5 cm) rectangle from paper. Lay the paper flat. Use a mountain fold to fold it in half lengthwise.

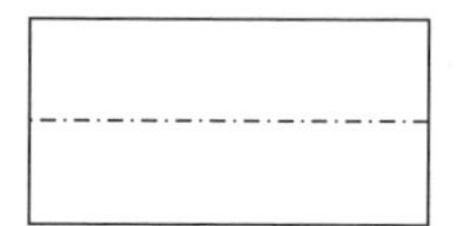

2. Measure the top point as shown, and draw diagonal lines to the corners. Then, with the paper folded in half as in step 1, sink-fold the corners. A sink fold changes part of a mountain fold into a valley fold.

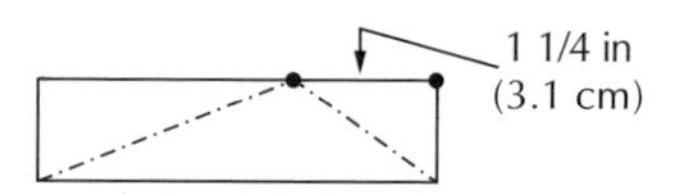

3. Press the canopy flat to finish it.

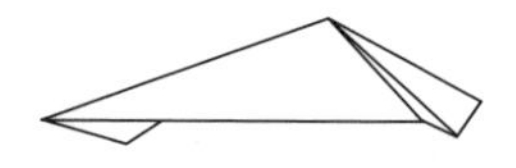

Step 6

Apply glue to the inside of the nose only, and insert the canopy. Align it with the nose. Adjust the shape so that the wings have a slight dihedral angle (upward slant) and the winglets slant upward, as shown.

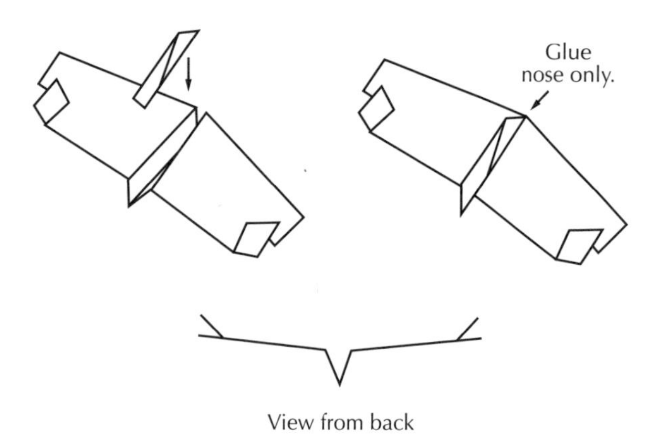

©2005 by The National Council of Teachers of Mathematics, Inc. www.nctm.org.
All rights reserved. For classroom use only. Written permission required for all other uses.

Tailless Planes to Soar This Summer

Experts hope models pave the way for stealthier jets

By Michael McCabe, Chronicle Peninsula Bureau

Two small-scale models of a tailless airplane that designers hope will lead to a jet fighter far more stealthy and agile than anything flying today will be flown this summer in Southern California.

The flight tests of the remotely piloted aircraft, dubbed the X–36 by the Pentagon, are also expected to show that it is possible to build a lighter jet fighter that can fly farther, officials at NASA's Ames Research Center in Mountain View and the McDonnell Douglas Corporation announced yesterday.

Airplane designers have long dreamed of a jet that could fly without a tail, which would lead to several exciting possibilities, said Larry Birckelbaw, NASA's X–36 program manager at Moffett Field. A tailless airplane means less weight and structure, which translates into far more maneuverability and less drag. It also means less surface to be detected by enemy radar.

"This is really a breakthrough in several technologies all developed over the past seven years right here at NASA's Ames and by a strong team from McDonnell Douglas," Birckelbaw said. "We hope this will be the plane of the future."

'The public has become so accustomed to a tail that they may have some discomfort climbing aboard a tailless aircraft'

—Larry Birckelbaw, of NASA

What makes the X–36 unique is not so much that it will be tailless—the B–2 Stealth bomber does not have a tail—but that it will be far more agile and less detectable, designers said.

One of the keys to the radical new design is that the direction of the nozzle of the main jet engine can be aimed in different directions, enabling the aircraft to fly more efficiently. The X–36 will be unveiled publicly on March 19 in a ceremony at McDonnell Douglas facilities in St. Louis.

The X–36 will be 18 feet long, have a wingspan of 10 feet and weigh 1,300 pounds. During test flights at the NASA Dryden Flight Research Center at Edwards Air Force Base the two aircraft will be powered by the same engine used in Cruise missiles. NASA officials refused to give more specifics, citing security reasons.

The combined cost of the program, to be roughly split by NASA and McDonnell Douglas, is $17 million for the development, fabrication and testing of the two aircraft.

Birckelbaw said it is too early to predict when the X–36 will evolve into a piloted aircraft, or even when tailless commercial airplanes may appear in the skies. But there are potential problems, mostly psychological. "The public," Birckelbaw said, "has become so accustomed to a tail that they may have some discomfort climbing aboard a tailless aircraft."

Used with permission of the *San Francisco Chronicle*, © 24 February 1996

From *Mission Mathematics II: Grades 6–8*, ©2005 by The National Council of Teachers of Mathematics, Inc. www.nctm.org. For classroom use only.

Test-Flight Data Sheet

Date: ______________________________

Mission Specialist's Name: ______________________________

Team Name: ______________________________

Members: ______________________________

____

____

____

____

Line Plots for Distance Trials:

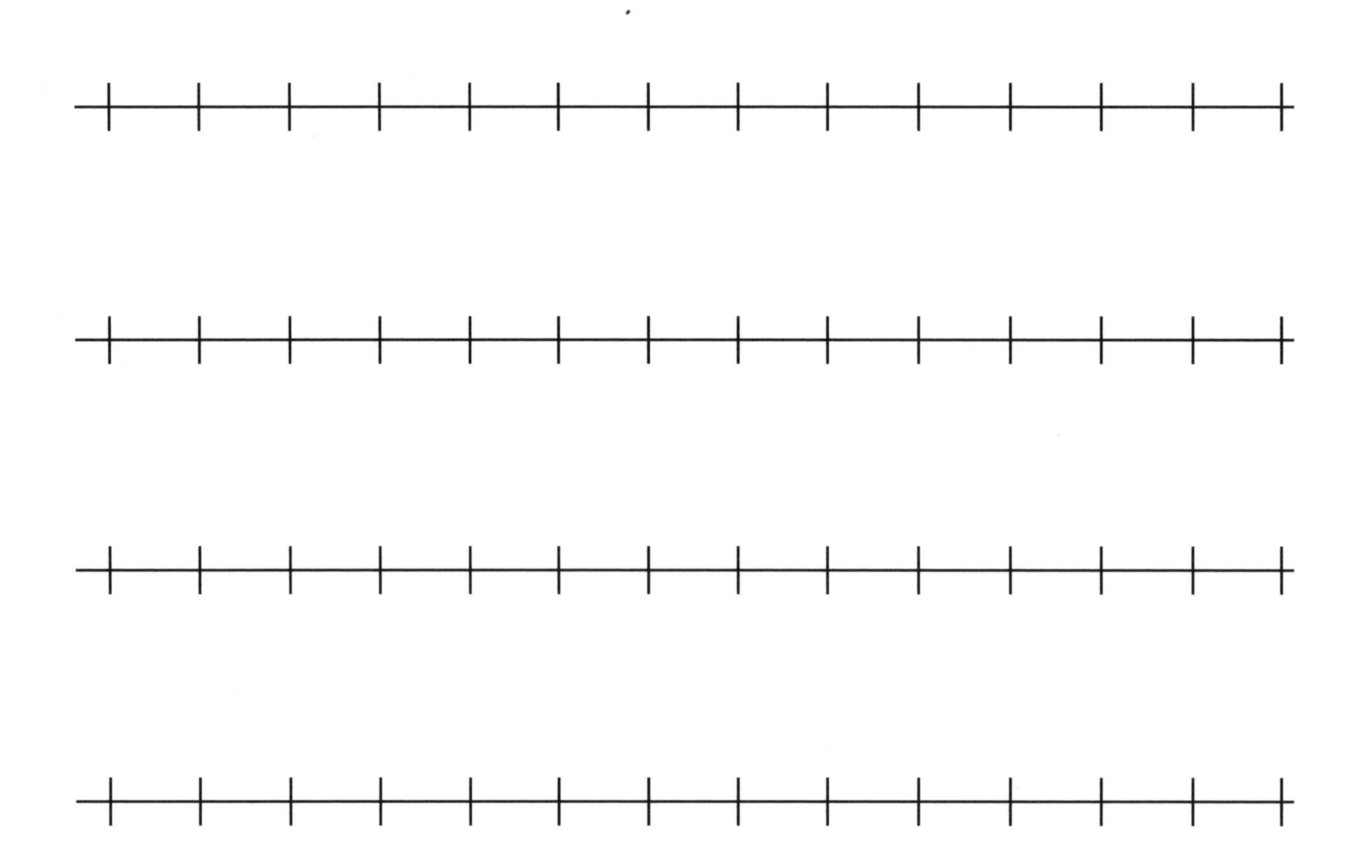

©2005 by The National Council of Teachers of Mathematics, Inc. www.nctm.org.
All rights reserved. For classroom use only. Written permission required for all other uses.

Event Specifications Form

Air Olympics
Event Specification Form

Event Title:

Expert Team Members:

Describe the event in words:

Describe the rules for the event:

Describe the judges who will be needed at this event and tell about their duties:

Describe the equipment that will be needed for this event:

Provide a scale drawing of the area needed for this event. Be sure to indicate the location of an area from which participants can safely watch, and also indicate the locations of the judge(s).

©2005 by The National Council of Teachers of Mathematics, Inc. www.nctm.org.
All rights reserved. For classroom use only. Written permission required for all other uses.

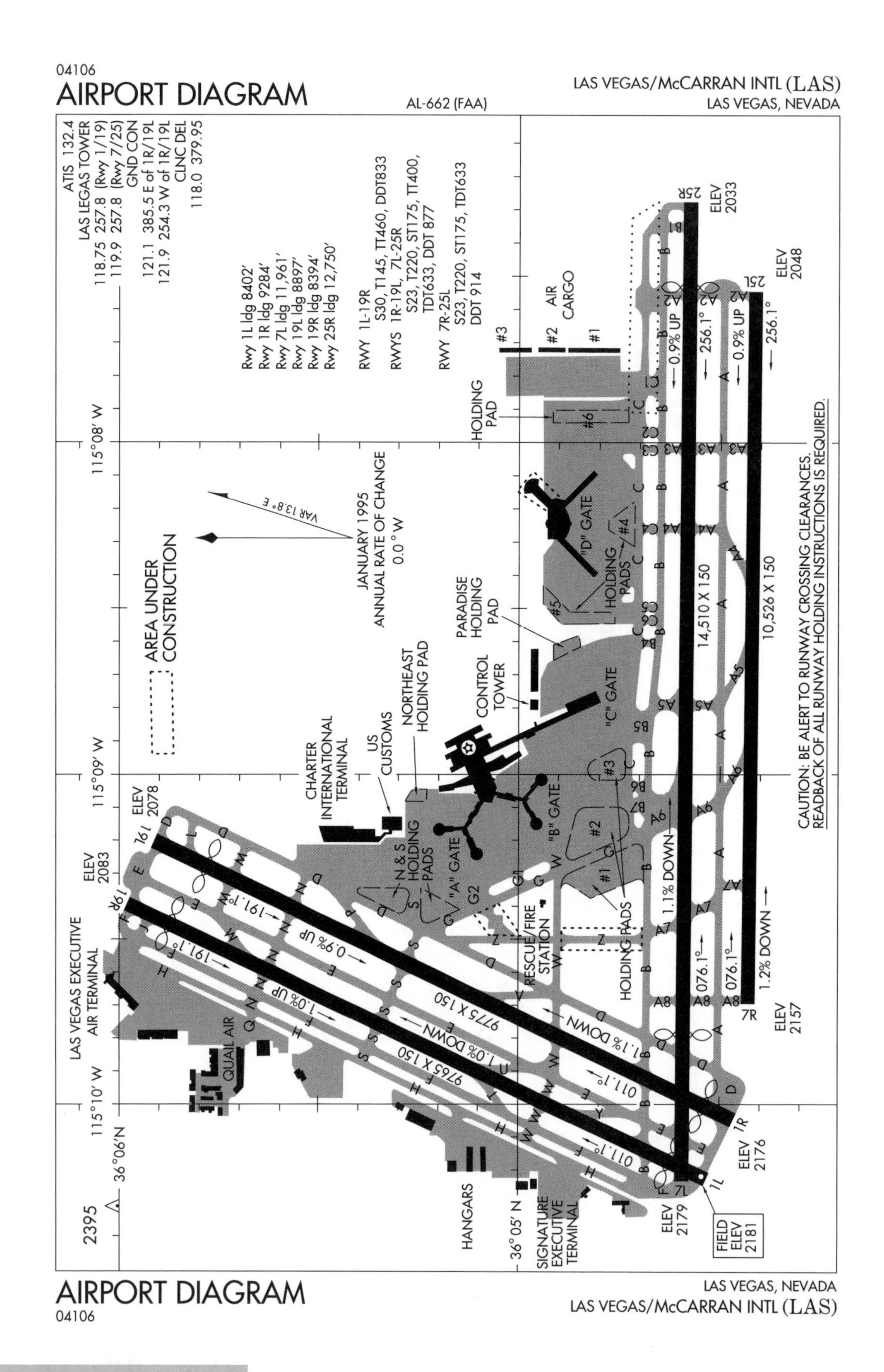

From *Mission Mathematics II: Grades 6–8,* ©2005 by The National Council of Teachers of Mathematics, Inc. www.nctm.org. For classroom use only.

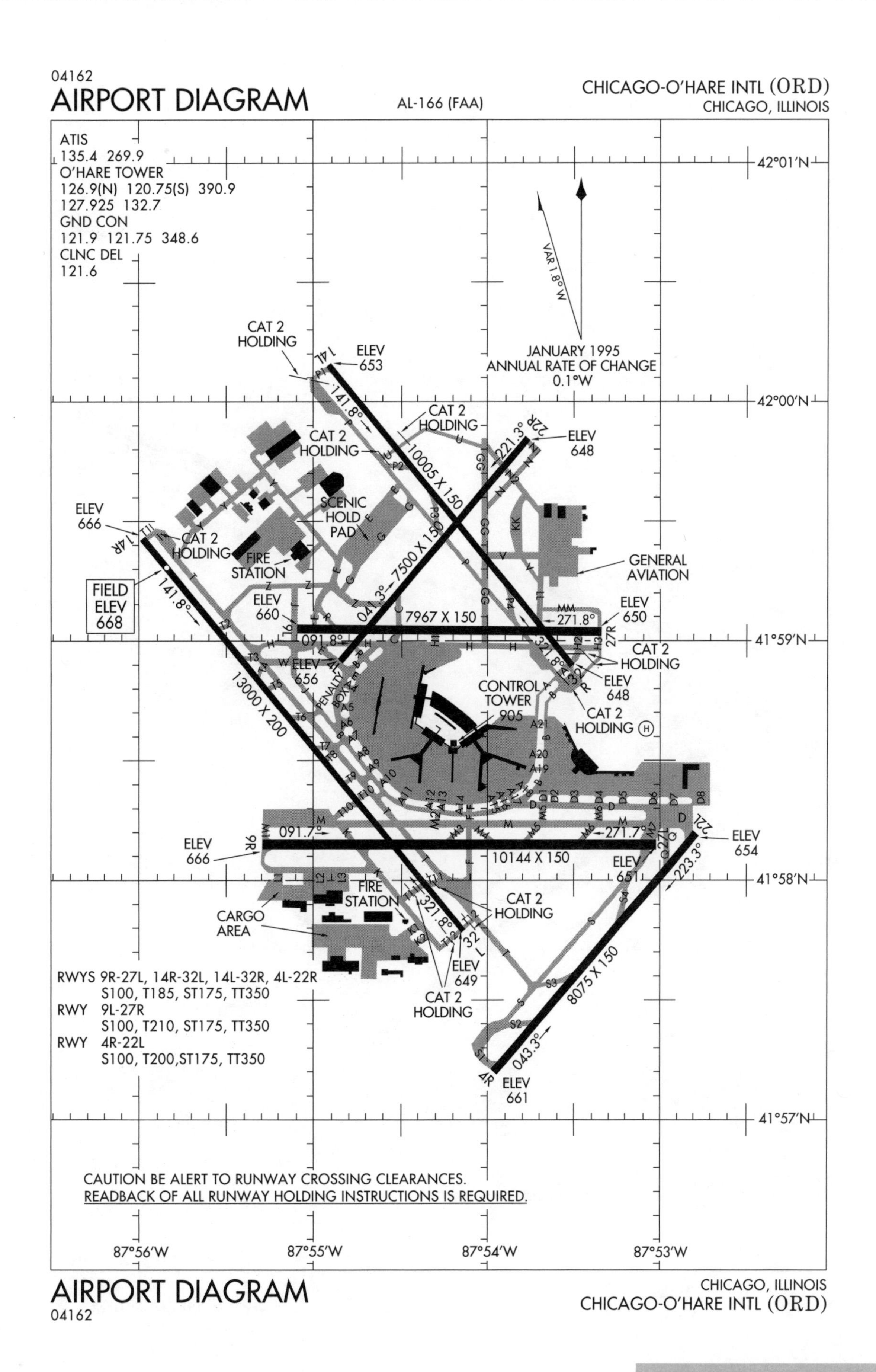

From *Mission Mathematics II: Grades 6–8*, ©2005 by The National Council of Teachers of Mathematics, Inc. www.nctm.org. For classroom use only.

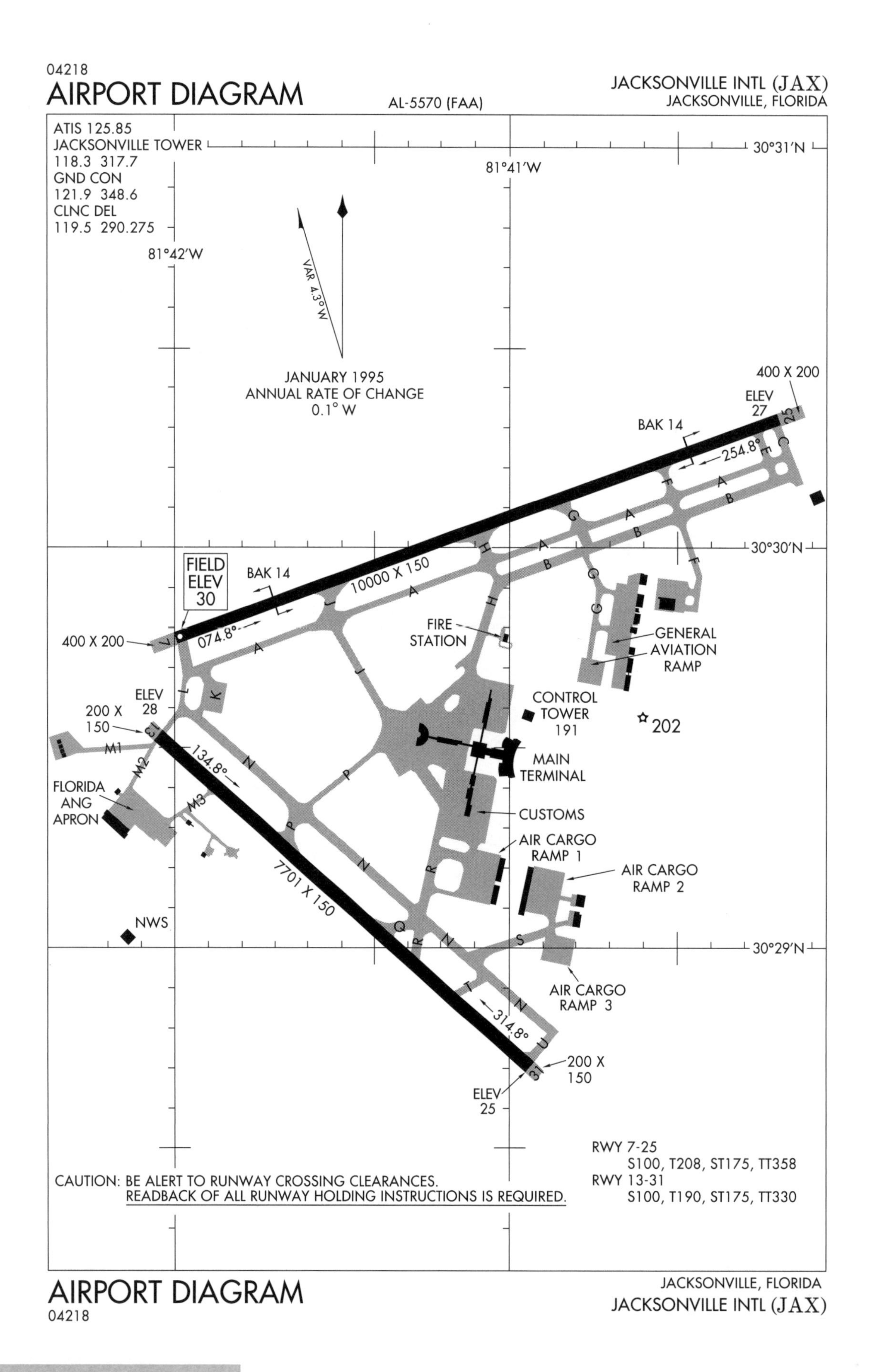

From *Mission Mathematics II: Grades 6–8,* ©2005 by The National Council of Teachers of Mathematics, Inc. www.nctm.org. For classroom use only.

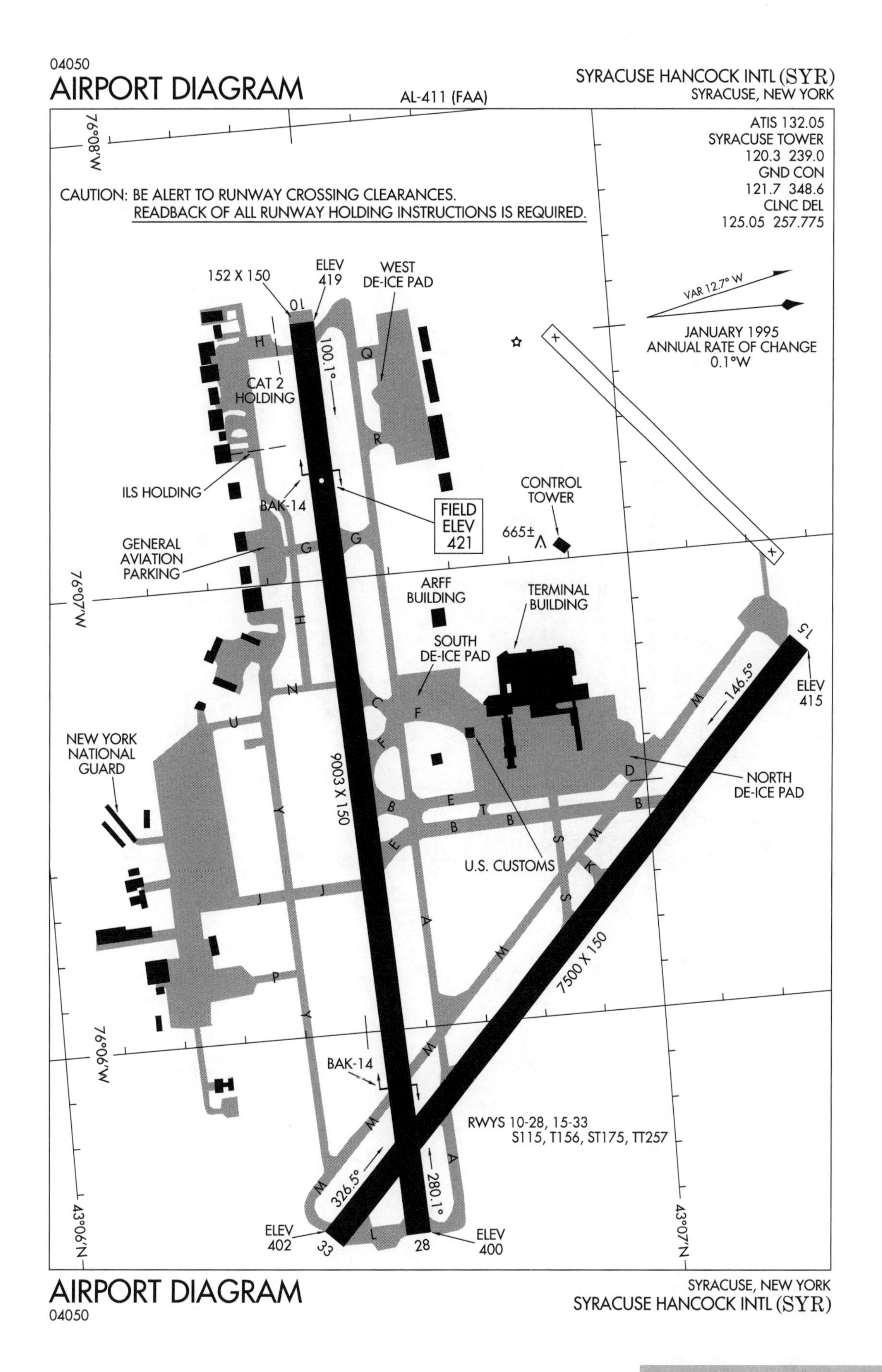

From *Mission Mathematics II: Grades 6–8*, ©2005 by The National Council of Teachers of Mathematics, Inc. www.nctm.org. For classroom use only.

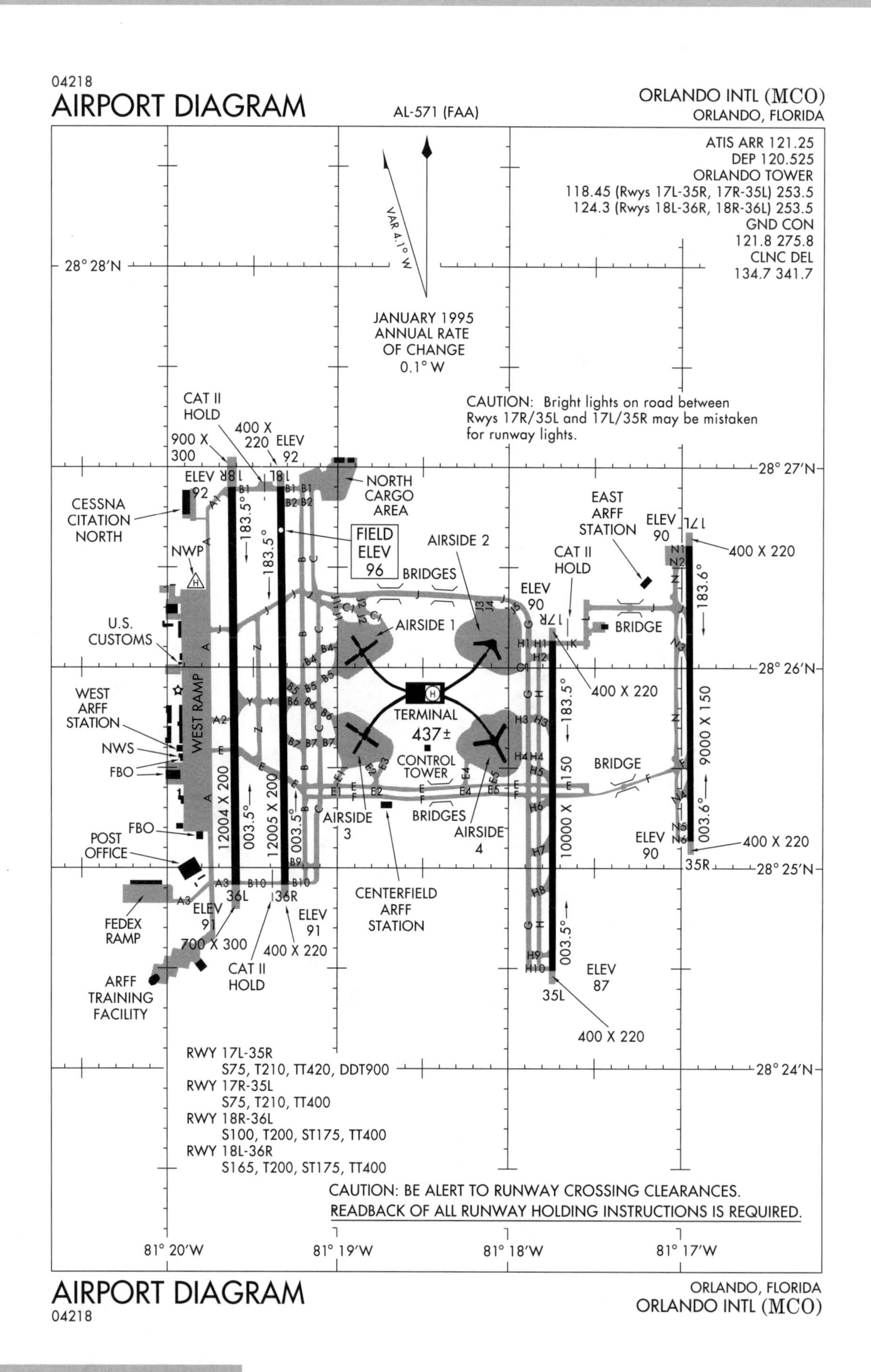

From *Mission Mathematics II: Grades 6–8*, ©2005 by The National Council of Teachers of Mathematics, Inc. www.nctm.org. For classroom use only.

Ice Mass

Determine the area of the ice mass represented by the scaled figure below.

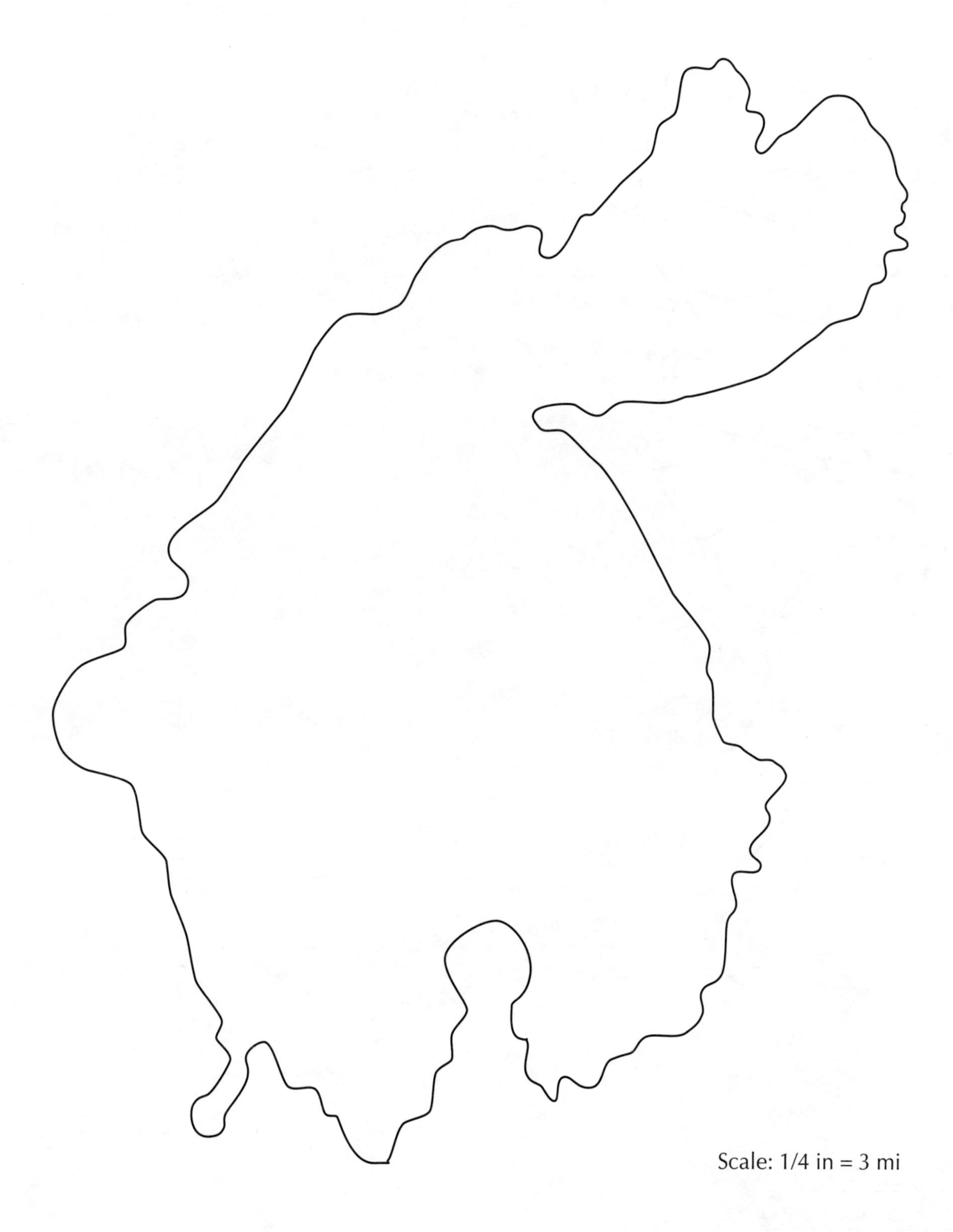

Scale: 1/4 in = 3 mi

©2005 by The National Council of Teachers of Mathematics, Inc. www.nctm.org.
All rights reserved. For classroom use only. Written permission required for all other uses.

U.S. East Coast and West Indies

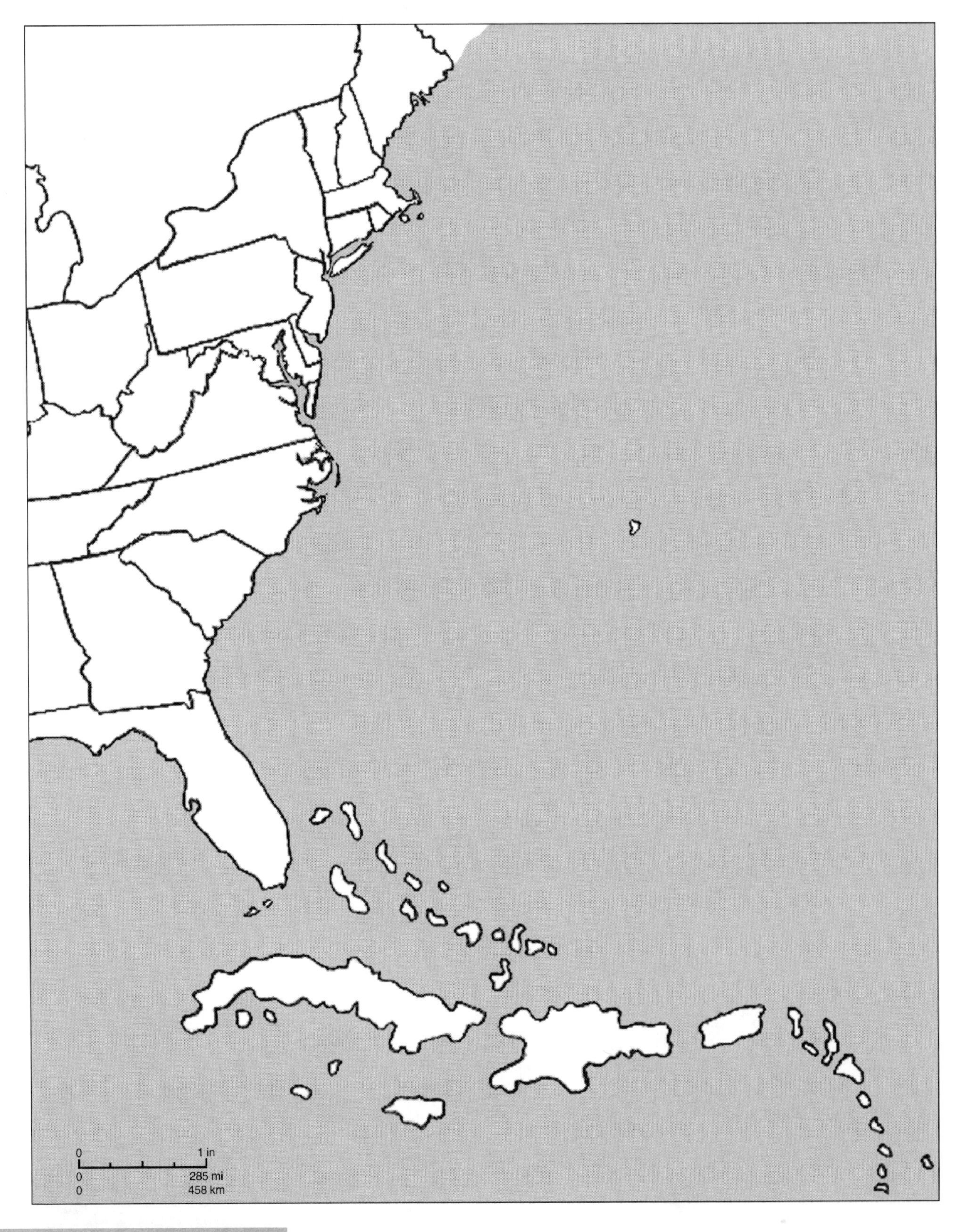

©2005 by The National Council of Teachers of Mathematics, Inc. www.nctm.org.
All rights reserved. For classroom use only. Written permission required for all other uses.

Thar They Blow

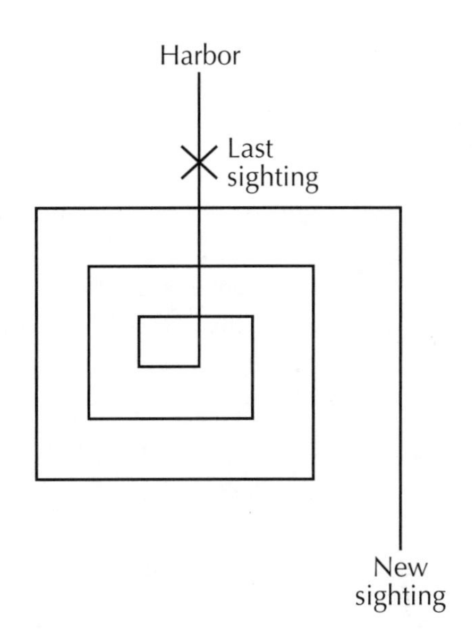

Captain Ahab, a Maine sea captain, has a whale-watching boat, the *Acadia*. He takes passengers out for half-day or whole-day excursions to see humpback whales. The key to his success is that Captain Ahab almost always finds whales for his passengers to see. He uses his knowledge of whale behavior and his loran (*lo*ng-*ra*nge *n*avigation) unit to locate whales. He listens to his radio, too. Often other boats will use a ship's radio to tell others of whales sighted.

On an excursion Captain Ahab uses the loran receiver to look for locations where he has seen whales recently. He knows the location of the dock. He uses this location and the location of a recent sighting to determine an initial course to follow. If he arrives at a loran-marked site and there are no whales, he searches in a pattern. He begins by sailing south for a specified distance, then turns west for a time, then north, and finally east. Of course, he cannot turn sharply enough to make a right-angle turn, but he tries his best to do so. He hopes to produce a search pattern of an ever widening "right angle" spiral (see the illustration at right).

One afternoon Captain Ahab went south from the harbor to look for the humpbacks where he had seen them on the morning excursion. Alas, the whales had moved. He began his usual search pattern. He had traveled 12 miles to the south from the point of the morning sighting when he heard a radio message from a boat that had seen the whales. After a conversation with the radio operator on the other boat, he found that they were due west of the *Acadia*. Captain Ahab turned the boat 90 degrees clockwise and proceeded to the location. They traveled 5 miles and located the whales. How far had the whales traveled since they were last seen by Captain Ahab?

Draw a picture of the course that Captain Ahab followed to find the whales, and use the data to calculate the straight-line distance the whales traveled.

©2005 by The National Council of Teachers of Mathematics, Inc. www.nctm.org.
All rights reserved. For classroom use only. Written permission required for all other uses.

Latitude and Longitude

Activity 1

Work in pairs. The first partner selects one of the locations below. Using the given latitude and longitude, the second partner points to the location on a map or globe and identifies the major city near it. The first partner then checks the work (the names of the cities are given at the bottom of the worksheet). When the second partner has identified two locations correctly, reverse roles. Repeat the exercise until all the locations have been identified.

1. 28° N, 97° W
   What Texas city is near this location?
2. 40° N, 105° W
   What Colorado city is near this location?
3. 42° N, 71° W
   What Massachusetts city is near this location?
4. 28° N, 82° W
   What Florida city is near this location?
5. 40° N, 120° W
   What Nevada city is near this location?
6. 35° N, 90° W
   What Tennessee city is near this location?
7. 21° N, 158° W
   What city is near this location?
8. 15° N, 121° E
   What capital city is near this location?
9. 35° S, 58° W
   What capital city is near this location?
10. 34° S, 18° E
    What seaport city is near this location?

Activity 2

Work in pairs. The first partner points to a city or an important landmark on a map or globe. The second partner identifies the latitude and longitude of the city or landmark. The first partner then checks the work and accepts answers that are within 5° of the actual latitude and longitude. Reverse roles after the second partner has correctly identified two locations. Repeat the exercise until all the locations have been identified.

The following are some examples of cities and landmarks:

1. Washington, D.C.
2. Hong Kong
3. The Eiffel Tower
4. The Kennedy Space Center
5. Cleveland, Ohio
6. Niagara Falls
7. Bar Harbor, Maine
8. Marquette, Michigan
9. Dallas, Texas
10. Phoenix, Arizona
11. Pierre, South Dakota
12. Victoria Falls (Africa)

Activity 3

Work in pairs. Use your map skills to answer the following questions:

1. Part of what state is north of 60° N?
2. What city in the mainland United States is closest to the equator?
3. What is the longitude of the easternmost point in the mainland United States?
4. What state reaches farther than 170° W?
5. Follow the 40° N latitude line from the East Coast of the United States to the West Coast. What major cities lie close to this latitude?

Answers

Activity 1: 1. Corpus Christi; 2. Denver; 3. Boston; 4. Orlando; 5. Reno; 6. Memphis; 7. Honolulu; 8. Manila; 9. Buenos Aires; 10. Cape Town. **Activity 2:** 1. 39°N, 77°W; 2. 22°N, 114°E; 3. 49°N, 2°E; 4. 28°N, 81°W; 5. 42°N, 82°W; 6. 43°N, 79°W; 7. 44°N, 68°W; 8. 47°N, 87°W; 9. 33°N, 97°W; 10. 33°N, 112°W; 11. 44°N, 100°W; 12. 17°S, 26°E. **Activity 3:** 1. Alaska; 2. Key West; 3. Approximately 67°W; 4. Alaska; 5. Some examples are Philadelphia, Pennsylvania; Columbus, Ohio; Indianapolis, Indiana; Springfield, Illinois; and Denver, Colorado.

©2005 by The National Council of Teachers of Mathematics, Inc. www.nctm.org.
All rights reserved. For classroom use only. Written permission required for all other uses.

GPS Receiver Position Screen

Position

315°
heading

60 mph
speed

1.2 mi
odometer

N 39° 45′ 00″
W 105° 00′ 00″
position

5280 ft
altitude

14:19:29
time

©2005 by The National Council of Teachers of Mathematics, Inc. www.nctm.org.
All rights reserved. For classroom use only. Written permission required for all other uses.

Compass

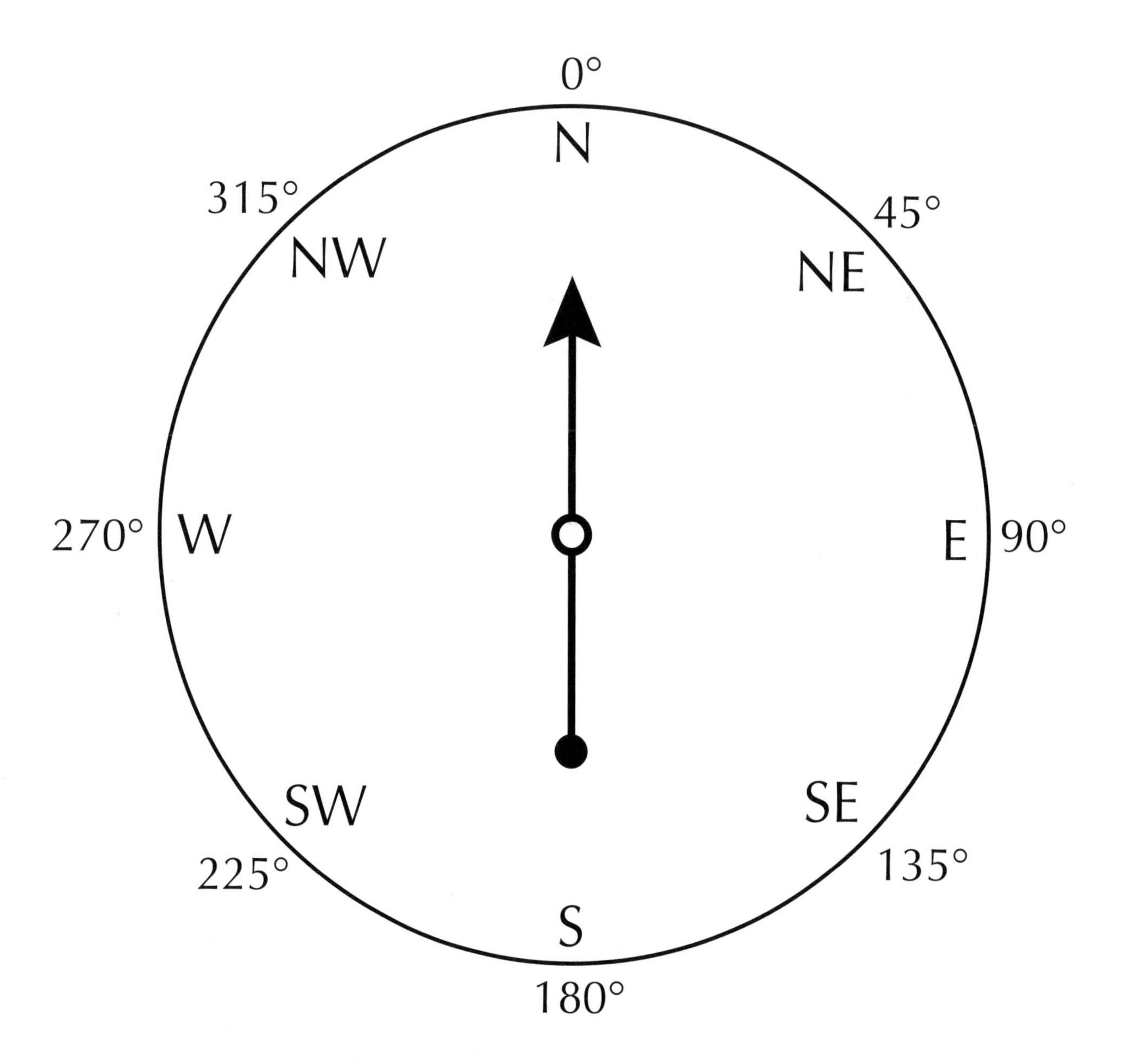

©2005 by The National Council of Teachers of Mathematics, Inc. www.nctm.org.
All rights reserved. For classroom use only. Written permission required for all other uses.

GPS Receiver Pointer Screen

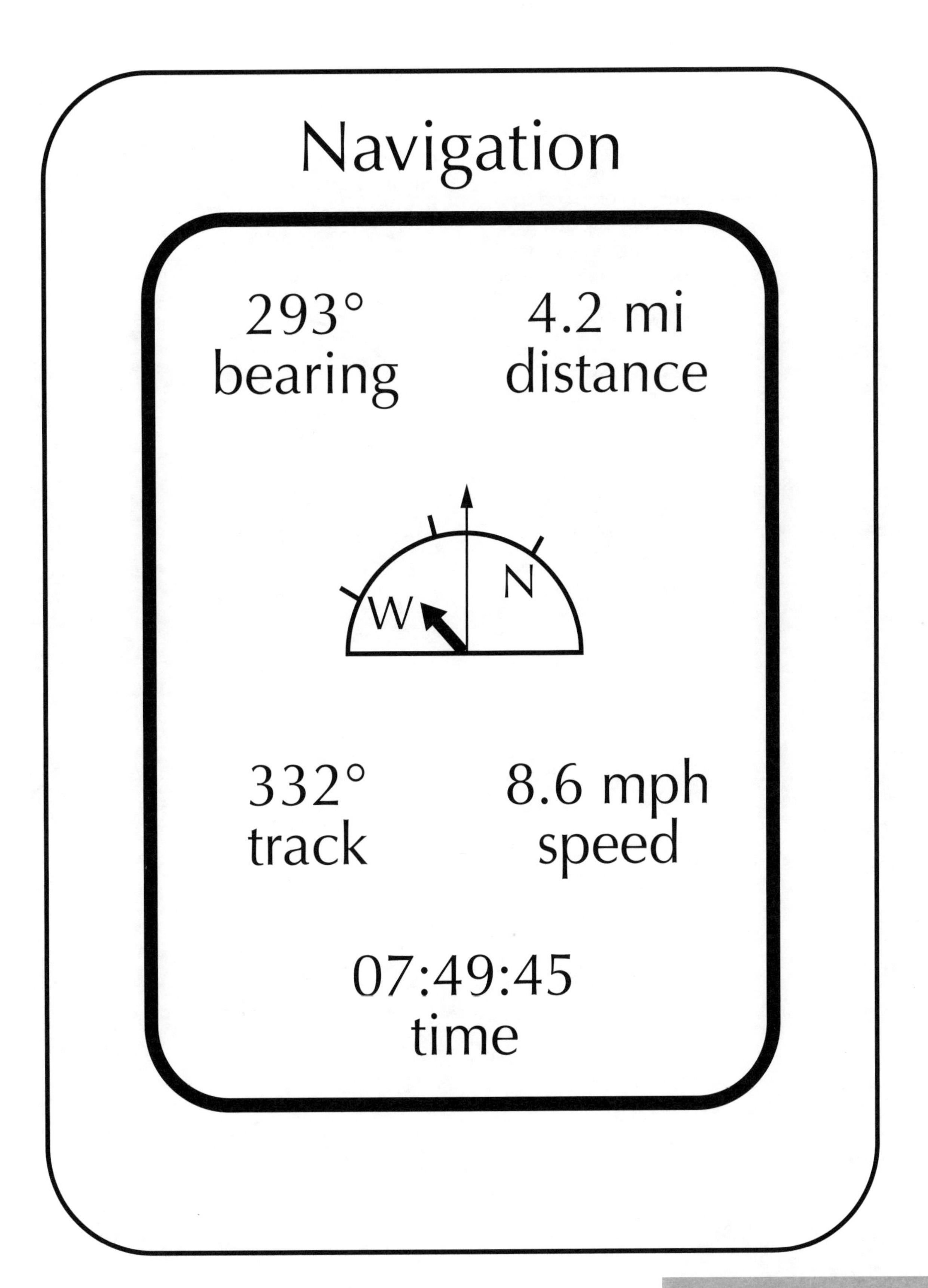

©2005 by The National Council of Teachers of Mathematics, Inc. www.nctm.org.
All rights reserved. For classroom use only. Written permission required for all other uses.

Bearings and Tracks

Mission Specialist's Name: ______________________

Use your paper-plate compass to help you set the following bearings on the pictures of compasses provided. Show your answer on a paper-plate compass to a classmate near you. When you agree that the answer is correct, draw the arrow on the illustration.

1. Bearing: 60°

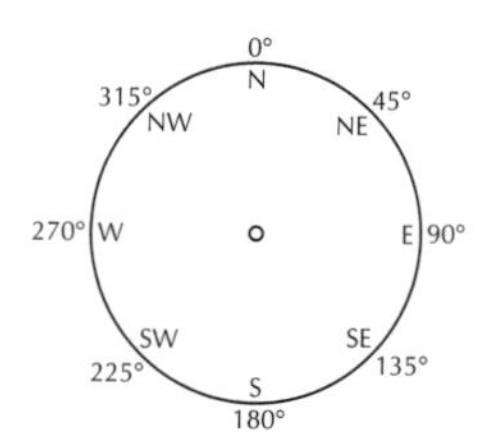

2. Bearing: 275°

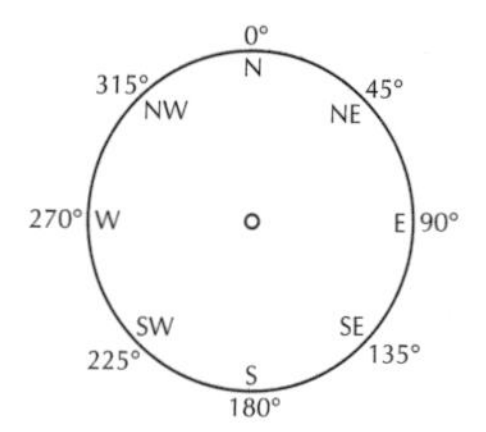

3. Bearing: 155°

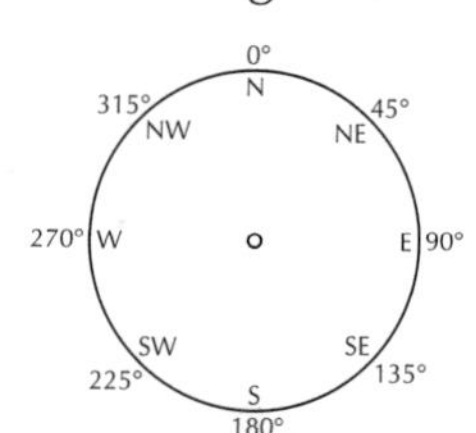

4. Bearing: 340°

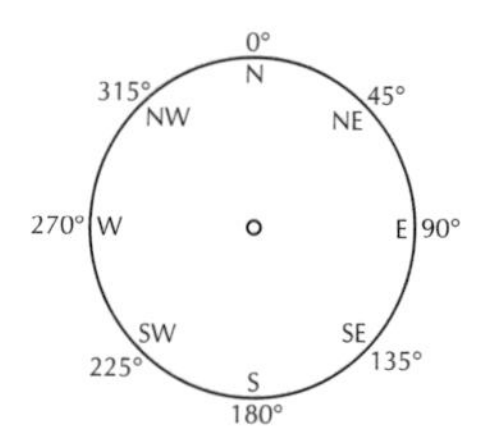

Below are several compasses with bearings shown. Estimate the number of degrees shown by each of the illustrations.

1. Bearing: _____

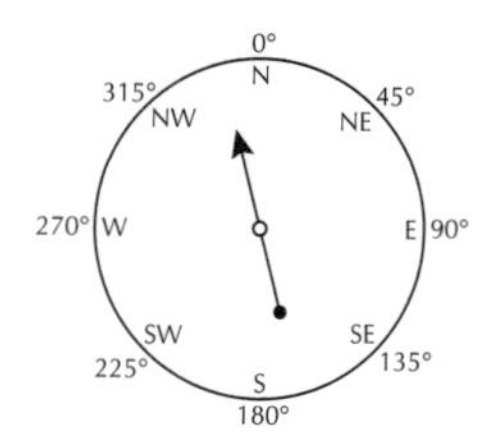

2. Bearing: _____

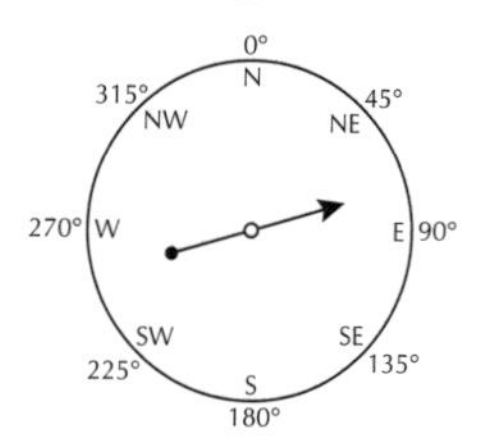

3. Bearing: _____

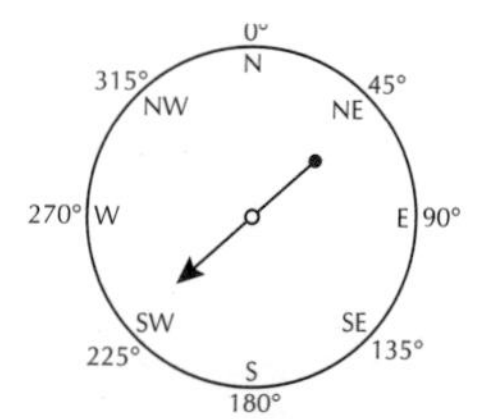

4. Bearing: _____

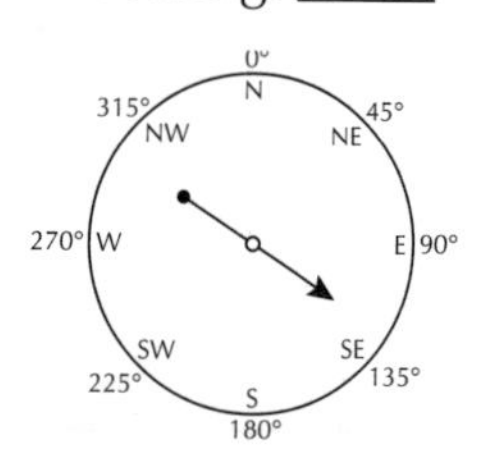

In the illustrations below, the track is shown by the longer arrow and the bearing is shown by the shorter arrow. Look at each illustration. Normally, ships' captains make the shortest turn to align the bearing and track. Tell whether you, the ship's captain, would turn left or right to make the track and bearing become the same.

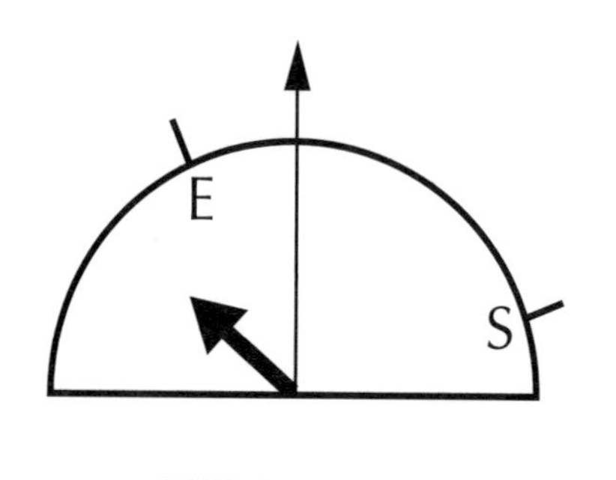

____

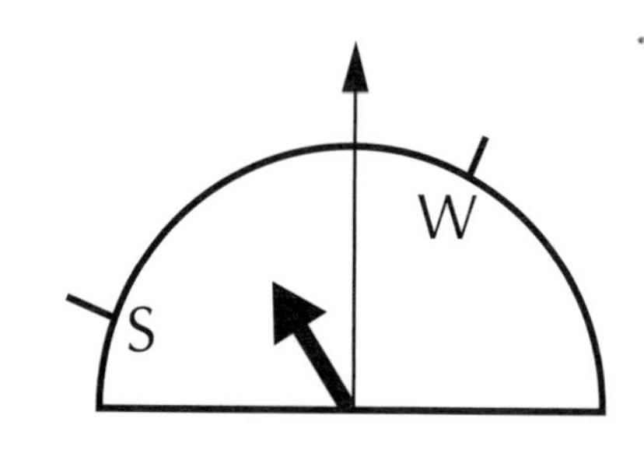

____

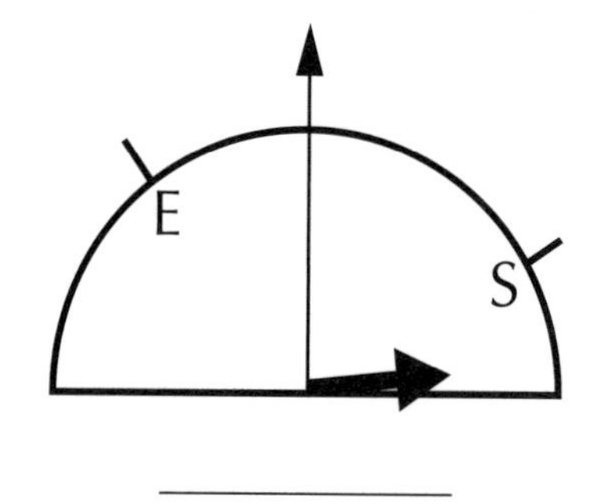

____

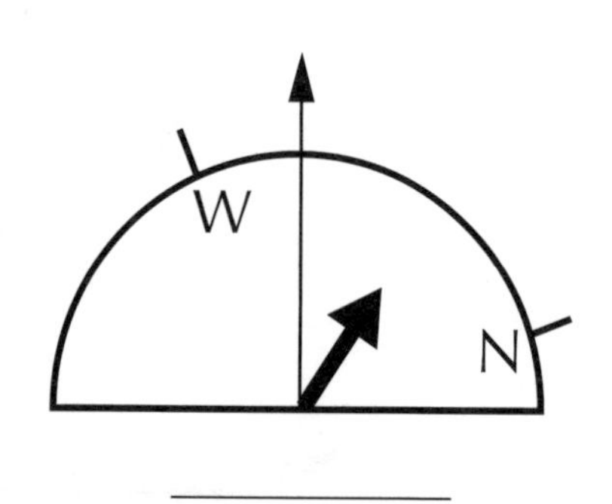

____

©2005 by The National Council of Teachers of Mathematics, Inc. www.nctm.org.
All rights reserved. For classroom use only. Written permission required for all other uses.

Bibliography

Adams, Thomas. "Letting Your Students 'Fly' in the Classroom." *Mathematics Teaching in the Middle School* 2 (March/April 1997): 348–49.

Belsky, Nancy Ann. "Math by the Month: Mission Mathematics by the Month." *Teaching Children Mathematics* 4 (March 1998): 408–9.

Berkman, Robert M. "Exploring Interplanetary Algebra to Understand Earthly Mathematics." *Teaching Children Mathematics* 5 (October 1998): 78–83.

Billstein, Rick. "You Are Cleared to Land." *Mathematics Teaching in the Middle School* 3 (May 1998): 452–57.

Blackburn, Ken, and Jeff Lammers. *The World Record Paper Airplane Book.* New York: Workman Publishing: 1994.

Cave, Linda. "Super-Jumbo Jet: An Airborne Village." *Mathematics Teaching in the Middle School* 3 (October 1997): 128–31.

Gay, David, and William Yslas Velez. "Living Graphs." *Mathematics Teaching in the Middle School* 7 (November 2001): 172–77.

House, Peggy A., and Roger P. Day, eds. *Mission Mathematics II: Grades 9–12.* Reston, Va.: National Council of Teachers of Mathematics, 2005.

House, Peggy, and Michael C. Hynes, eds. *Mission Mathematics: Grades 9–12.* Reston, Va.: National Council of Teachers of Mathematics, 1997.

Hynes, Mary Ellen, ed. *Mission Mathematics: Grades K–6.* Reston, Va.: National Council of Teachers of Mathematics, 1997.

Hynes, Mary Ellen, and Donn Hicks, eds. *Mission Mathematics II: Grades 3–5.* Reston, Va.: National Council of Teachers of Mathematics, 2005.

Hynes, Mary Ellen, Juli K. Dixon, and Thomasenia L. Adams. "Rubber Band Rockets." *Teaching Children Mathematics* 8 (March 2002): 390–96.

McCabe, Michael. "Tailless Planes to Soar This Summer." *San Francisco Chronicle,* February, 1996. sfgate.com/cgi-bin/article.cgi?file=/chronicle/archive/1996/02/24/MN70367.DTL.

National Aeronautics and Space Administration (NASA). *Spinoff 1994.* Washington, D.C.: NASA, 1994.

———. *Spinoff 1995.* Washington, D.C.: NASA, 1995.

National Council of Teachers of Mathematics (NCTM). *Curriculum and Evaluation Standards for School Mathematics.* Reston, Va.: NCTM, 1989.

———. *Professional Standards for School Mathematics.* Reston, Va.: NCTM, 1991.

———. *Assessment Standards for School Mathematics.* Reston, Va.: NCTM, 1995.

———. *Principles and Standards for School Mathematics.* Reston, Va.: NCTM, 2000.

Nord, Gail D., and David Jabon. "The Mathematics of the Global Positioning System." *Mathematics Teacher* 90 (September 1997): 455–60.

Piper, Watty. *The Little Engine That Could.* New York: Platt & Munk Publishers, 1999.

Schmidt, Norman. *Best Ever Paper Airplanes.* New York: Sterling Publishing, 1994.

Whitin, David J. "Mathworlds." *Teaching Children Mathematics* 3 (March 1997): 417–18.

Young, Mark C., vol. ed. *Guinness Book of World Records.* Stamford, Conn.: Guinness Media, 1996.

THREE ADDITIONAL TITLES APPEAR IN THE

Mission Mathematics II

Series

(Michael C. Hynes, Project Director)

Mission Mathematics II: Prekindergarten–Grade 2
Edited by Mary Ellen Hynes and Catherine Blair

Mission Mathematics II: Grades 3–5
Edited by Mary Ellen Hynes and Donn Hicks

Mission Mathematics II: Grades 9–12
Edited by Peggy A. House and Roger P. Day

Please consult
www.nctm.org/catalog
for the availability of these titles,
as well as for a plethora of
resources for teachers of mathematics
at all grade levels.

For the most up-to-date listing of NCTM resources on topics of interest to mathematics educators, as well as information on membership benefits, conferences, and workshops, visit the NCTM Web site at www.nctm.org.